Arid Land Irrigation and Ecological Management

Arid Land Irrigation and Ecological Management

Nehal Kidwai

RANDOM PUBLICATIONS
NEW DELHI - 110 002 (INDIA)

Arid Land Irrigation and Ecological Management

ISBN 978-93-51112-75-4

Published in 2014 in India by

Reprint 2020

RANDOM PUBLICATIONS

4376-A/4B, Gali Murari Lal, Ansari Road
New Delhi-110 002
Phone: +9111-43580356, 23289044
E-mail: randomexports@gmail.com; sales@randompublications.com; info@randompublications.com

Type Setting by: Friends Media, Delhi-110089
Printed at : Mehra Printers, Delhi-110092

Preface

The irrigated agriculture which utilize river or groundwater are often introduced to arid lands because of the scarcity of annual rainfall. In case of Pakistan in West Asia, irrigation facilities which can utilize abundant water of Indus river have been constructed since the colonial period and vast areas have been irrigated. However, excess irrigation has been causing problems such as water logging and salt accumulation in the soil. Similar problems are reported in Syria as well. In UAE in the Arabian Peninsula, agricultural and/or afforestation projects which seems more or less environmentally unconscious have been implemented with abundant oil money. These projects are causing deterioration of water quality, drying up of groundwater and salt accumulation in the soil due to inadequate irrigation with saline water. Thus, it is urgently needed to plan and implement appropriate projects in terms of scale and level with nationwide viewpoint. As mentioned above, problems like salt accumulation are often caused by irrigated agriculture in arid lands, and two methods seem to be effective to avoid those problems. One is to irrigate adequate amount of water, and the other is to drain the excess water away. However, when the scale of irrigation facility or organization is very large, it is rather difficult to control the amount of water provided to each farmer properly and timely. As a result, farmers are apt to irrigate excessively when water is available. This is understandable because, in arid lands, crops easily die without irrigation and grow better when irrigated with more water. As long as such situation continues, it would be almost meaningless if agriculture extension officers and experts of aid organizations intend to train farmers how and when to be irrigated. These problems can be solved if farmers can use enough water whenever they want, which means irrigation facilities should not be too large so that farmers can control them by themselves.

When irrigated agriculture is newly introduced to arid lands, irrigation using well or river water is generally adopted. But, before the modern irrigated agriculture, usually they have traditional cultivation practices even in such arid areas. "Water harvesting" which makes use of scarce rainfall is one of them. And "Oasis agriculture" which often uses underground canals is another traditional way in arid lands. Compared to the modern agriculture, crop yield with traditional way is unstable and poor. However, in terms of sustainability, traditional practices are more advantageous than the modern ones which tend to cause problems such as salt accumulation within a relatively short period. On the contrary, traditional practices have been used for a long time and causing few problems, which proves that the practices suit the climate and other conditions of the areas and conserve the environment. Thus, when we engage agricultural development in arid lands, it is indispensable to study traditional cultivation practices carefully.

The book will be an indispensable source for all professionals, researchers and students in this subject and for anyone working in the related areas for acquiring an up-to-date overviews.

I thank all members of my team who have helped in the preparation of the book. My special thanks go to "Random Publications" who have published the book.

—Nehal Kidwai

Contents

1

Surface and Groundwater Resources of Arid Zone

Water is a precious resource that is currently in need of long-term planning for storage and judicious use for the survival of mankind. Excessive exploitation of water without any systematic conservation programme can lead to depletion of the resource beyond recovery, as is being currently experienced in the arid regions. The water resources of the two most problematic arid areas in the country, the hot arid areas of western Rajasthan and the cold arid areas of Ladakh district in Jammu and Kashmir.

Water Resources of Arid Western Rajasthan

Western part of Rajasthan state is the most problematic area within the hot arid zone. The average annual rainfall here is 318 mm, as compared to 531 mm in Rajasthan state. Due to low and erratic rainfall, replenishment of exploited water is also very poor. During the Twentieth Century, different parts of western Rajasthan experienced agricultural drought once in three years to every alternate year. The overall probability of drought for the state is 47%. The weather condition even in average years for most part of the year remains too dry and inhospitable for successful growth of crops. Under such conditions of uncertainty, conventional cropping is risky and has necessitated widespread use of groundwater for irrigated cropping. Although the whole of Rajasthan state is categorized as water-scarce (having per capita water availability below 1000 m^3 year-1 Narain *et al.*, 2006a), the condition in western Rajasthan is more precarious. The west-central part of western Rajasthan is devoid of any drainage network and has meager surface water resources, which adds to the problem. Rapid urbanization and industrialization, increasing pollution of the water sources by the industrial units and over-extraction of water from deep wells has led to water quality problems. Despite climatic adversity, this is one of the most densely populated deserts of the world.

Long-term statistical analysis of rainfall data of the region indicates an asymmetric average storm intensity profile for storms of short duration, with the highest intensities falling in the first part of the storm. Occurrence of water flow in the watercourses is unpredictable, of short duration and high variability. The instantaneous discharge-duration curve shows very high and irregular peaks indicating the problems of controlling runoff. Regulation of sporadic discharge by means of surface reservoir presents many problems, especially the high ratio of storage to mean annual runoff volume required to produce the degree of control necessary for economical agricultural development (Goyal and Vittal, 2008).

Hydrological Zones

Hydrologically western Rajasthan has been divided into three broad zones (Venkateswarlu *et al.*, 1990).

Zone – I : Region with major input of surface water from more humid region, frequently with extensive irrigated agriculture. About 60% area of Ganganagar district in the north and 50% area of Bikaner district and 25% area of Jaisalmer district in the northwest lie in this zone. This is the main canal irrigated zone in arid Rajasthan.

Zone-II: Plain lands with a primitive or no stream network. The region has a system of repetitive micro-hydrology. Churu, Jhunjhunu, Sikar, Nagaur, Jodhpur and parts of Bikaner, Jaisalmer and Barmer districts come under this category. This zone occupies 52% area of arid Rajasthan.

Zone-III: Sloping region with an integrated stream network. The Luni basin, occupying the districts of Pali, Jalore, part of Jodhpur and Barmer districts, lie in this zone.

Surface Water Resources

The inherent surface water resources of the western Rajasthan are scarce and because of low and erratic rainfall, replenishment of these water resources is also very poor. Due to high atmospheric temperature and low humidity, a large part of the rainwater is lost as evapotranspiration. Except in canal command area in north, surface water potential is very low in the central, western and southern parts.

More than 50% area of northwest Indian arid zone comprises of sandy plains, dune systems, eroded rocky/gravely surfaces and isolated hillocks. In central and western parts, the run-off generated in response to some high-magnitude rainstorms gets lost in sandy terrain. As per initial estimate approximately 280 x 106 m3 surface water is available annually for utilization in this part (Jain, 1968). Mehta and Kashyap (1970) estimated the surface water potential of this region as 200 x 106 m3, out of which 130 x 106 m3 is utilisable resource.

Sharma and Vangani (1992) estimated the surface water potential of this region as 1360 x 106 m3 out of which 47% was utilized till 1988. Nearly 33% area of the zone is occupied by the Luni Basin, the Sahibi Basin and a few smaller river basins. These are ephemeral drainage systems and convey runoff only in response to torrential rainfall during the monsoon season. The estimates of surface water resources for Luni basin ranges from 518 x 106 m3 (Dhir and Krishnamurthy, 1952), to 571 x 106 m3 (Dhruvanarayana *et al.*, 1964), to 868 x106 m3 (Mehta, 1970), to 939 x 106 m3 (Mehta and Kashyap, 1970), to 858 x 106 m3 (Anon., 1980), to 1130 x 106 m3 (Sharma, 1991). Large numbers of tanks, reservoirs, minor irrigation dams and check dams have been constructed at different locations in Luni basin and other areas to store runoff water during monsoon. About 550 storage tanks in the capacity ranging from less than 1.51 to 208 x 106 m3 are functional with total utilizable capacity of nearly 1169.28 x 106 m3 for providing irrigation to 0.102 x 106 ha land (Khan, 1997). Out of these, six reservoirs viz. Jaswantsagar, Sardar Samand, Jawai, Hemawas, Ora and Bankali, are the major irrigation tanks with capacity of irrigation of more than 4000 ha each. Jawai is the main source of drinking water supply to many towns and villages (Source: Irrigation Department, Govt. of Rajasthan).

About 15% area in the north-west arid zone of India receives major input of water through an extensive canal network. This region comprises of the arid districts within the states of Haryana and Punjab and the northwestern part of Rajasthan. The canal systems in this region are the Gang Canal, the Bhakra Canal and the Indira Gandhi Canal bringing water by diverting the flows of the Sutlej, the Beas and the Ravi in Punjab. The estimated designed capacity of these canals varies from 314 to 524 cumec (Murthy and Gulati, 1978; Uppal, 1978; Gupta, 1987; Kapoor and Rajvanshi, 1977). On the basis of average flow from 1982 to 1996 the canal water available in the four districts of Ganganagar, Hanumangarh, Bikaner and Jaisalmer is at the rate of 0.90 m ha-1 (Anon., 1988; 1996), far more than the stipulated mean value of 0.51 m ha-1. Indira Gandhi Nahar Pariyojna (IGNP) The Indira Gandhi Nahar Pariyojna (IGNP) receives its supplies from the Indian share of waters of Indus river basin as per Indus Water Treaty of September 1960 between India and Pakistan.

The three eastern rivers of the basin (namely Ravi, Beas and Sutlej) have been harnessed since then through several facilities dams, reservoirs, links and canals. Although water resources development of the Indus waters dates back to centuries, major activities started in the Tenth Century and continued to meet with the growing needs of the agricultural community. The Gang canal system built during 1922-1929 utilizes 1.11 million acre feet (MAF) of water which constitute the pre-partition share of India, drawn from the Sutlej river near Ferozpur Head Works.

RD 45 of the Gang canal was linked with Harike Barrage constructed in the upstream at a later date. The inter-state agreement for sharing waters of the Ravi-Beas rivers came into being in the year 1955 between erstwhile State of Punjab, Jammu and Kashmir (J&K) and Rajasthan, followed by Bhakra Nangal agreement in 1950. Agreements were later revised and finally as per 1981 agreement total share of water for Rajasthan in Ravi-Beas waters was 9.71 MAF. This quantum of water was further allocated by the Government of Rajasthan to IGNP stage I and II, 3.59 MAF(inclusive of 0.22 MAF for drinking and other uses) and 4.00 MAF (inclusive of 0.65 MAF for drinking and other uses), respectively. The remaining water was given to Gang Canal 1.44 MAF (1.11+0.33), additional water for Bhakra system from Ravi-Beas waters 0.21 MAF and Sidhmukh/ Nohar Project 0.47 MAF.

Stage I consists of a 204 km long feeder canal, having a headworks discharge capacity of 460 m3 sec-1, which starts from Harike Barrage. 170 km of the feeder canal lie in Punjab and Haryana and 34 km in Rajasthan. The Stage-I also consists of 189 km long main canal and 3454 km long distribution system, which are concrete lined, and serve 553 kha of culturable command area, out of which 46 kha are served by pumping to a 60 m lift, through four pumping stations. IGNP Stage II comprises onstruction of a 256 km long main canal and 5,606 km of a lined distribution system, and will serve 1,410 kha of CCA (873577 ha area in flow and 537018 ha under lift), utilising 4,930 Mm3 yr-1 of water. The main canal in the entire length was completed in the year 1986. The irrigation intensity initially planned for Stage I and Stage II was 110% and 80%, respectively (GOR, 1999). The canal system was initially designed at a water allowance of 5.23, 3.0

and 2.0 cusecs/1000 acres for Stage I, Stage II flow area and Stage II lift areas, respectively (which works out to 0.36, 0.21 and 0.14 cumecs per 1000 hectares).

Irrigation in canal command area needs proper management to avoid water logging, soil salinity, etc. More than 60% of canal command area has sandy soils with poor water holding capacity. Nearly 50% of additional water applied in form of irrigation goes as deep percolation and joins groundwater. The problem of water logging is more apparent in recent years. The mean rate of water table rise varies from 1.1 m year-1 in Stage-I to 0.81 to 0.85 m year-1 in Ghaggar plain and Bhakra command, respectively, and 0.64 m year- 1 in Gang canal command area. On the basis of annual rise in water table, it has been found that an area of 1456 km2 has already turned critical (water table within 6 m of land surface). However, a far more serious water logging problem awaits Stage-II, owing to an underground hard substratum of gypsum within 10 m depth. In about 34% area (1205 km2) of the gross command area of 3544 km2, water collected in low lying areas does not seep down (Rahmani and Soni, 1997). Due to capillary action, the water comes to the surface with dissolved salts and evaporates, leaving the salt behind, thus making the land saline. According to one study (Chouhan, 1988), if surface drainage is not introduced in the waterlogged area, thousands of hectares of land will be submerged and salinised in 25 to 30 years.

In the adjoining Haryana state, Sirsa and Hisar districts have 148 km2 (3.0%) and 266 km2 (4.0%) area, respectively, within the influence of high water table (Rao *et al.*, 1986); the rate of water table rise is in the range of 0.14 to 1.0 m year-1. In addition to salts contributed by groundwater nearly 2.0 x 106 t of salts are added annually through canal irrigation. In Punjab state nearly 1.05 x 106 ha area is irrigated through canals in the arid districts of Faridkot, Firozpur and Bhatinda (Sidhu *et al.*, 1991). As a result about 80% of the total salt affected and waterlogged area of the state occurs in these districts; the water table is rising @ 0.52 to 0.75 m year-1.

Groundwater Resources of Arid Rajasthan

Groundwater resources of this region are very poor in terms of quantity and quality. Groundwater in this region is not sufficient even for drinking purposes. Over and above insufficient quantity, the groundwater is moderately to highly saline over large area. A dominantly sandy terrain and disorganized drainage network (drainage density is as low as 0.3 km km-2), and recurring droughts constantly exert pressure on already meager groundwater resources. The stage of groundwater development has exceeded 100% in Barmer, Jalore, Jhunjhunu, Jodhpur, Nagaur, and Sikar districts. Number of Safe blocks has been significantly reduced because of meager rainfall and over exploitation of groundwater resources mainly for irrigation.

Groundwater Depletion

Due to over mining groundwater levels are declining in 9 out of 12 districts of arid Rajasthan since 1984. Groundwater table in Jalore and Pali districts shows a decline rate of more than 0.50 m year-1. In Jodhpur, Jhunjhunu, Nagaur and Sikar districts groundwater decline rate is 0.44-0.48 m year-1. In Barmer, Churu and Jaisalmer districts the rate of decline is less than 0.20 m year-1. In Bikaner, Ganganagar and Hanumangarh districts water level shows a rising trend due to IGNP.

Future Groundwater Scenario

Future projection of groundwater utilization has been worked out for the year 2010, 2015, 2020 and 2025 considering the average growth rate @ 3.20% compounded annually, though the present actual growth rate is higher.

The projection presents a grim situation even when the numbers are on a lower side. The stage of development is likely to reach 161.6, 189.1, 221.4 and 259.1% in the year 2010, 2015, 2020 and 2025, respectively. Arid Rajasthan has to face the adverse effects of overexploitation. Strategies for Water Resource Management in Arid Zone Rain is the principal source of water in this region, which augments soil moisture, groundwater and surface flows. Agriculture and several other economic activities in arid areas depend on rain. Of the total water use about 85% of water is used for irrigation and remaining 15% is used for drinking, industrial and other purposes. About 65% of irrigation water and 30-40% of drinking water is subjected to serious losses. Hence, increasing water use efficiency, coupled with increasing availability of water through rainwater harvesting and management, is key to survival on sustainable basis. Rainwater harvesting, its conservation and efficient utilization can solve problem of water scarcity to a greater extent. Rainwater harvesting in small ponds (*nadis*), underground tanks (*tankas*), *Khadins* (Low lying areas) etc., is an age-old tradition here. These traditional structures vary in design, shape and size. These structures are now partially or sometimes totally neglected because of increased dependence on tubewell, tankers, canal water supply, etc. The traditional methods require revival and improvement for being more economical and efficient.

Rainwater Harvesting in Nadis and Ponds

The people of rural arid areas live in scattered settlements called *dhani's* distributed over sand dunes, interdunal plains and undulating landforms. Under such conditions it is inconceivable that organized water supply will be feasible to fully meet the demand of thirsty land, human and livestock. Rainwater harvesting in *Nadi* and farm pond for groups of farm families or community are the most viable proposition. *Nadi* is a dugout pond used for storing runoff water from adjoining natural catchment during rainy season. High evaporation and seepage losses through porous sides and bottom, heavy sedimentation due to biotic interference in the catchment and contamination are its major bottlenecks. Complete control of seepage and evaporation losses is very costly and not foolproof. Nadi can play a significant role in preventing complete crop failure (Mann and Singh, 1977; Singh, 1983). Singh (1986) reported better utilization of nadi water for raising nurseries, orchards and to support the initial establishment of trees observed instead of watering field crops.

Chatterji *et al.* (1985) reported pollution of nadi water due to free access of human and livestock, leading to the growth of many harmful bacteria and other water-borne diseases and therefore, not safe for human consumption in Nagaur district. Sedimentation in pond is another major problem. Shankarnarayan and Singh (1979) observed reduction in water surface area and drainage basin area up to 1.8 to 2.4 and 6 to 8 Times, respectively, due to biotic interference.

To overcome these problems CAZRI has developed designed *Nadis* with LDPE lining on sides and bottom keeping surface to volume ratio 0.28 and provision of silt trap at inlet (Khan, 1989). *Nadis* also help in recharging groundwater aquifers although their effect varies depending on the underlying soils and rocks. Where the substrate is rocky, it is estimated that they contribute a depth of 0.06 metres of water a year c ompared to 1.58 metres in sandy plains. A study of a 2.25-hectare *nadi* with a storage capacity of 15,000 cubic metres in the north Gujarat alluvial area calculated that the pond contributed as much as 10,000 cubic metres of water to the groundwater aquifer in one rainy season.

Farm Pond is an improved version of *nadi* with treated catchment and surplusing arrangement for removal of excess water. A farm pond of 20,000 m3 capacity was constructed at Kukma watershed at Bhuj in Gujarat by CAZRI in year 2004. Construction of farm pond resulted in assured availability of 20,000 m3water even in during 150 mm rainfall (Narain *et al.,* 2006 b). The collected water was used to provide irrigation to Datepalm, ber, aonla and other fruits plants in nearby area. Construction of large number of rainwater harvesting nadis and farm ponds can solve the problem of uncertainty of occurrence of rainfall and can store water during heavy rainfall for non-monsoon period for human, livestock and crops on sustainable basis. Therefore, construction/renovation and desilting of nadis/farm ponds during drought relief measures by state government and NGO's can be beneficial.

Rainwater Harvesting in Khadin for Crop Production

Recurring droughts and long dry spells are regular feature of arid zone of Rajasthan, which result in crop failure or severe reduction in crop growth and yield. A traditional practice of *khadin* farming in arid Jaisalmer district ensures better moisture conservation and cropping. The system is very effective even where annual average rainfall is less than 200 mm. Khadin farms were first constructed by Paliwal Brahmin of Jaisalmer district in the Fifteenth Century (Sehgal, 1973). The Paliwals connected most of the local catchments into well-knit system of *khadin* farms for assured crop production even under low rainfall. Kolarkar *et al.* (1983) observed 2 to 16 times lower electrical conductivity (EC) of Khadin soils compared to outside farms. The reduction in EC has been attributed to leaching of salts through seepage water in khadin. The *khadin* soils hold soil moisture to last up to growing season. The average yield of 20 to 30 q ha-1 for wheat and 13 to 25 q ha-1 for chick pea without any specific agronomical practices and fertilizers were also reported under khadin. Tiwari (1988) has linked the *khadin* cultivation with the farming system of Kalibanga and Rang Mahal cultures of the Indian desert, which are presumed to be of Harappan age (3000 BC).

CAZRI has evolved a design package and guidelines for construction of khadins (Khan, 1992 a). Improved khadin has been constructed by CAZRI near village Danta in Barmer district. The catchment area of the khadin is 137 ha with 6.88 ha submergence. Provision of 40 m bed bar in 450 m long earthen embankment was provided for spilling over excess water in khadin bed. The total water storage capacity of khadin is 54.2 x 104 m3 and beneficiaries are four farm families (Khan, 1998 a). A Khadin of 20 ha area was developed in Baorali-Bambore watershed with surplussing arrangement. Before construction of Khadin, uncontrolled runoff from upper catchment used to wash away seeds, fertilizers, and standing crops, besides loss of valuable water. After construction of Khadin, farmers could take excellent Kharif and Rabi crops (Narain and Goyal, 2005). Collecting water in a khadin aids the continuous recharge of groundwater aquifers. Studies of groundwater recharge through khadins in different morphological settings suggest that 11 to 48% of the stored water contributed to groundwater in a single season. This replenishment of aquifers means that subsurface water can be extracted through bore wells dug downstream of the khadin. The average water-level rise in wells bored into sandstone and deep alluvium was 0.8 metres and 1.1 metres, respectively. There were 500 such *khadins* covering an area of 12,140 ha in Jaisalmer and the crop production from such areas was adequate to feed the people of Jaisalmer district. Largescale development of *khadin* farms at suitable locations in western Rajasthan can enhance land productivity.

Rainwater Harvesting from Constructed Catchment

Under this system, catchments are constructed in such a way that runoff is directed to a desired destination instead of spreading here and there. The desired destination may be crop/tree to supplement the soil moisture or a reservoir for storage and subsequent utilization. Inter-row, inter-plot and micro-catchment are some of the constructed catchments, which can be used to enhance the availability of water. Runoff from microcatchments was generally found to depend upon rainfall and catchment characteristics.

The results of seven years' field studies showed that microcatchments produced 13 to 32% of rainfall as runoff at 0.5% slope; and even higher amounts at 5% (36 to 45%) and 10% (26 to 44%) slopes. Runoff generally increased with decreasing slope length; runoff for 5 and 10% slope were nearly equal but greater than for 0.5% slope (Sharma, 1986). Apparently, there is a critical slope beyond which runoff is not affected by slope increases. A large part of the rainfall is absorbed by the sandy soil of the catchment, thereby reducing the total amount of harvested water. Various surface covers and sealants were experimented to increase runoff. Plastic covered catchments were found to generate 95% runoff while Janta emulsion (asphalt), pond sediment and compacted catchments yielded 91%, 88% and 66% runoff, respectively (Sharma *et al.*, 1986). The results of studies at farmer's field in Kalyanpur (Barmer district) showed very high efficiency of moisture conservation with stone and sand filled polythene bags with associated improvement in growth and establishment of Ziziphus mauritiana (ber) seedlings (Ojasvi *et al.*, 1999). Inter-row water harvesting using ridge-furrow has also been found useful in raising dryland crops.

Under this technique, 50-60 cm wide ridges alternated with 30-40 cm inside furrows (15 cm deep) are constructed using ridger equipment. Crops are planted in furrows, adopting a paired row design. Ridges yield runoff to the furrows, thus enhancing the moisture regime in the root zone. Singh *et al.* (1973) reported 210% increase in the yield of pearl millet with this system. They concluded that ridge-furrow technique has better adaptability for small holders, as no area of the field is lost to catchment construction. Singh (1982) suggested maintaining the furrows and ridges as permanent structures.

Thus, if the tillage is restricted to furrows only, the energy input can be substantially reduced. Rainwater harvesting for groundwater recharge As per groundwater estimates of Rajasthan for year 2001, total annual groundwater availability is 11,159 mcm as against total annual water demand (draft) of 11,626 mcm. The overall stage of groundwater development for Rajasthan state is 104%, which is categorized as 'overexploited'. Presently out of total 32 districts, 14 districts are in the category of overexploited, 4 in critical zone, 8 in semi-critical zone and remaining 6 are considered in safe category. With increasing demand for water, more and more blocks are likely to be in the category of overexploited and need immediate attention for recharge of groundwater. Percolation tanks, pondage in stock tanks with infiltration galleries, sand filled dam, anicuts across stream, sub-surface barriers, etc., are used for groundwater recharge (Ojasvi *et al.*, 1996; Goyal, 1999). Sub-surface barriers constructed across ephemeral streams trap sub-surface flow to recharge groundwater aquifer (Khan, 1998 b). Construction of two subsurface barriers of 10 m length each within 300 m from water supply well was found enough to store runoff water required for a village having a population of 500 (Anon., 1974). Singh *et al.* (1989) reported that soil conservation practices have increased recharge to the extent of 14.02 to 19.52% of rainfall in Udaipur region. Adoption of conservation measures like anicuts, loose stone check dams, brushwood check dam, etc., in Jhanwar watershed (Jodhpur district) resulted in recharge/increase of groundwater level @ 0.33 m-.75 m year-1 (Bhati *et al.*, 1997; Goyal *et al.*, 2007). In another watershed at Osian-Bigmi (1991-96), conservation measures like loose stone check dams, vegetative barriers and anicuts resulted in rise in water table by 1.1 m, indicating the effectiveness of conservation measures for recharge of groundwater (Gupta *et al.*, 2002).

Rainwater Harvesting through Tanka for Potable Water

Good quality potable water is a global issue, particularly in developing world because 80% of the diseases in the world is due to poor quality drinking water. The problem of poor quality groundwater used for drinking is more acute in Rajasthan. Concentration of fluoride ranges from 0.4 to 90 mg l-1 leading to various diseases like dental fluorosis, skeletal fluorosis and, non-skeletal manifestation etc.

Rainwater is the purest form of water. Appropriate harvesting of rainwater from roof top and open and its utilization can alleviate problem of fluoride to great extent. Studies conducted at CAZRI have revealed that roof made of different materials can generate 50- 80% runoff that can be stored in underground cistern (*tanka*) which could provide excellent drinking water round the year. Rainwater can also be harvested in tanka using artificially prepared catchment. Collection and storage of excess rainwater in tankas is an age-old tradition for meeting domestic water requirements. Till today most villages depend on these structures as source of drinking water. Traditional *tankas,* constructed with lime plaster, typically have a life span of three to four years. They suffer from seepage and evaporation losses and in the absence of proper silt traps and pollutant-free inlets, the quality of the conserved water deteriorates over time, making it unsafe for drinking. In many situations, degradation of the catchment area means that it does not yield the quantity of water required to continuously replenish the structure.

Increasing Water Productivity by Reduction in Water Losses

Evaporation accounts for over two third of water losses from surface water bodies in hot arid regions. Reducing surface area by increasing storage depth can appreciably reduce water losses. The surface area can also be reduced by storing the water in a compartmented reservoir, and pumping the water from one compartment to another as the water is used, so that there are some full compartments and some empty, instead of a single shallow sheet when the reservoir is partly used. Covering the field with any kind of mulch also helps in reduction of evaporation losses from the surface. Evaporation loss can also be minimized by reducing wind velocity through shelter-belts of suitable tree species around water bodies or by artificially shading of water surfaces. Studies on artificial shading of water surface have shown encouraging results in controlling evaporation losses. Shading of water surface with polyethylene sheet successfully reduced evaporation by 91%. Evaporation reduction with floating materials ranged from 37% for *Saccharum munja* to 82% for polystyrene sheet. Foamed rubber sheet, polyethylene sheet and bamboo reduced evaporation by 74%, 66% and 53%, respectively. The floating polystyrene sheet and polyethylene covers were the most economical (Khan *et al.*, 1990). The micro-irrigation systems like drip and sprinkler economize both water and fertilizers. These systems could be popularized to increase the productivity of limited rainwater.

Flash Flood Management

Although most parts of arid Rajasthan are drought-prone, flash floods are not uncommon in this region. Since rainfall in this region is of convective nature and usually occurs at a very high intensity for shorter duration creating flash flood, particularly in urban areas where buildings and roads generate very high runoff even for little rainfall. During 1979 large parts of state witnessed flash flood due to very heavy downpour of more than 500 mm in just 3-5 days which cut off the state from rest of the country for several days. Between 1901 and 2003 western Rajasthan witnessed nine moderate and 19 severe floods. Although floods are considered a natural calamity, however, if the excess flood water or its part can be managed and utilized for rejuvenating the depleted aquifers as well as improving surface water resources the problem of water scarcity during droughts can be solved.

Harvesting and conservation of floodwater to rejuvenate the depleted aquifers by adopting artificial recharge techniques will remarkably improve the water availability for growing population. Out of 12 district of arid Rajasthan, Bikaner, Hanumangarh and Ganganagar districts have reported a rise of groundwater table due to canal irrigation and remaining 9 districts have recharge potential of 30.3x109 cum based on groundwater depletion taken between 1984-2003. Narain *et al.* (2006 a) have estimated a potential of 5.9 x 109 cum of flash flood for western Rajasthan. A part of flash flood can be used to recharge groundwater of these districts. Bilara limestone, Lathi Sandstone, Jodhpur Sandstone and Alluvium aquifers are most suitable for groundwater recharge.

Watershed Management for Soil and Water Conservation

Initial approach of conservation programs was based on particular land problem and individual holdings. In this approach often the upper reaches of the fields were left untreated and runoff from uplands used to damage the conservation work laid out in the individual holdings at downlands. The watershed approach helps in an integrated development of different parts of the watershed in accordance with their nature, problems and potential with best amount of interference. The integrated watershed management principles have accordingly been adopted for holistic development of drought prone and rainfed areas. In areas where there is no defined drainage system and internal drainage is choked in sand dune complex, the 'Index Catchment Approach' was considered for taking development activities. Index catchment is demarcated by considering the crest line of sand dunes or eroded rocky outcrops from the line of water divide (Vangani *et al.*, 1998). Focused efforts on watershed management were started after 1983 when Govt. of India (GOI) launched model watershed projects with involvement of Indian Council of Agricultural Research. Under this program 47 model watersheds were identified in different agro-climatic zones all over the country. Rajasthan got its first model watershed during 1986-87 as Jhanwar model watershed near Jodhpur. Later on, to cover large areas of dryland, the National Watershed Development Program for Rainfed Area (NWDPRA) was launched in 1990-91 by GOI. This scheme covered blocks with less than 30% area under irrigation. Under this scheme Rajasthan initiated development in 204 watersheds in 190 Panchayat Samities. To involve large segments of the rural community in this venture, new guidelines for watershed development were adopted from 1st April 1995, and subsequently revised in August 2001. In 2008 National Rainfed Area Authority issued common guidelines for watershed management that became effective from 1st April 2008.

Lessons learnt in watershed management

In Rajasthan out of 26.5 million ha (rainfed) requiring natural resource management through watershed development and alternative income generation activities so far only 4.6 million ha has been treated under various schemes. Area available for watershed treatment in western Rajasthan is 2.0 million ha, whereas so far only 0.274 million ha has been treated under various schemes (Govt of Rajasthan, 2006). One of the examples of the benefits of watershed development in degraded arid fringe is in the degraded Aravallis at Siha, Rewari (in Haryana), where the number of electrified wells have increased from 67 to 205, sprinkler set from 2 to 70 and irrigated area from 260 to 420 ha.

The livestock population increased from 882 to 1396. Consequently, milk production increased from 2997 to 5724 liters day-1 and overall income increased by 400 to 500% (Singh *et al.*, 1996). With proper soil and water conservation measures in watershed in the Aravallis, soil loss was reduced from 150 t ha-1 to less than 5 t ha-1. Development of water resources resulted in the increase of net irrigated area to 28 ha and cropping intensity from 128 to 210%. Food grain and fodder production also increased substantially (Bhardwaj and Dogra, 1997). Likewise, adoption of graded bunds, gully control structures, contour cultivation, intercropping, use of cover crop in rotation along with other improved package of practices proved sustainable in the semi-arid region of Rajasthan. Graded bund has reduced the run-off from 20 to 4.8% and soil loss from 24 to 4.12 t ha-1 y-1. Besides other benefits, intercropping on contour resulted in 48% higher grain yield (Singh *et al.*, 1997). By adoption of various development activities in Osian index catchment during a five-year period, cropping intensity increased by 31.4% and forage yield by 1.97 t ha-1. Construction of water harvesting structures helped to increase the groundwater recharge as indicated by rise in static water level. Sediment deposition against loose stone check dam was 3.86 m/ha/yr (Vangani *et al.*, 1998).

For *in-situ* rainwater management circular micro-catchment of 5% inward slope with LDPE lining was successfully demonstrated in watershed area for establishment of *Ber* and other trees. For severely eroded and gullied catchment loose stone check dams (LSCD) were constructed at 1 m V.I. in Jhanwar watershed on 17 gullies, which proved very effective in controlling further extension of gullies and stabilized all the gullies (Goyal *et al.*, 2007). For channel treatment three masonry anicuts and two loose stone anicuts were constructed on the main streams at Jhanwar and Baorali- Bambore watersheds, respectively, which resulted in substantial reduction in water velocity and erosion downstream. Temporary inundation upstream helped in regeneration of vegetation beside recharging of groundwater. In Sar watershed, artificial recharge of groundwater was superimposed in a 2.8 ha m pond with three infiltration wells to improve water availability for conjunctive uses. For moisture conservation soil, straw and plastic mulch were tried in Baorali-Bambore watershed. The grain yield of pearl millet was 32.67% and 28.12% higher for plastic and straw mulch, respectively, over no mulch.

In alternative land use system various systems like agro-horticulture with pearl millet/mung/moth + Ber/Aonla/pomegranate, silvi-pastoral system with *C. ciliaris* + *Prosopis cineraria*/ *Colophospermum mopane*/*Harwickia binata* were successfully established in watershed areas. Ditch-cum-mound fencing and cut and carry system was adopted for pasture development in Jhanwar watershed (Bhati and Goyal, 1997). For wastelands alternative crops like *Cassia angustifolia* and *Lawsonia alba* were successfully raised at appropriate locations in the watershed. For arable farming improved varieties of pearl millet, clusterbean, mung bean, moth bean, etc., were introduced.

Water Demand under Global Warming

As per general circulation models (GCMs) with different scenarios of greenhouse gas emission, the globally averaged surface temperature is projected to increase by 1.4 to 5.8oC between 1990 and 2100. The possible change in temperature due to global warming will have serious implications for the present ecosystem. The greatest threat will be increase in evaporative losses and water demands caused by higher temperature (Minitzer, 1993). Globally evapotranspiration trends are projected for +5% to +10% increase due to increased temperature by +2°C to +5°C under equivalent doubling of atmospheric CO_2 from pre-industrial level (Schneider *et al.,* 1990). Wetherald and Manabe (1981) found that global evaporation changes by 3% when temperature changes by 1°C. Similarly, Budyko (1982) suggested 5% increase in evapotranspiration demand for each degree Celsius rise in temperature. Enhanced evapotranspiration would be primarily a consequence of higher air and land surface temperature.

Even in tropics, where temperature increases are expected to be smaller than elsewhere, the increased rate of moisture loss from plants and soil could be considerable (Rind *et al.*, 1989; Parry, 1990). Lal and Chander (1993) suggested a rise in annual mean surface temperature by 2.0-3.5°C over the Indian subcontinent by the year 2090. According to them warming would be most pronounced over northwestern India. The normal average annual evapotranspiration of Rajasthan is estimated as 1701 mm.

A small increase of 1% in temperature (£0.42°C based on normal maximum temperature of Rajasthan) will enhance the evapotranspiration demand by 11.7 mm on annual basis. It will cause an additional annual water demand of 718 mcm and 2250 mcm for the whole state based on net irrigated area (61345 Km2) and total cropped area (192302 Km2), respectively (Goyal, 2004). Total available utilizable groundwater for Rajasthan is 11159 mcm and the increase of 1% in temperature will put additional stress of 6.43% to 20.16% on existing groundwater resources and will reduce the number of safe districts from 6 to 3, bringing additional districts in the category of 'critical' and 'overexploited'. An increase in temperature by 2-3% from normal (i.e. 0.82-1.24oC) will leave only 1 district in the category of 'safe' zone. The remaining 31 districts will be mostly in the category of 'overexploited'. The satellite data of Rajasthan shows a total wetland area of 3450 Km2 which includes 1239 Km2 ha as natural and 2210 Km2 as man-made. Increase in evaporation due to global warming will cause additional annual water loss of 40.4, 80.7 and 121 mcm for 1, 2 and 3% increase in temperature, respectively. Globally, projected increase in temperature/evapotranspiration demand is coupled with increase in precipitation by almost same magnitude. However, it is projected that shift will be more towards extreme events. The area with higher rainfall will likely have more rainfall and areas with less rainfall will have lesser rainfall.

Since this state is not blessed with good perennial river systems, any increase in water demand requires careful planning for future water resource development. More emphasis is needed on development of technologies for reducing water losses in reservoirs, conservation of rainwater and development of such crop varieties that require less water.

Water Resources of Ladakh

In India the risk-prone high altitude areas include the easternmost trans- Himalayan part of Ladakh and Zanskar range (Jammu and Kashmir) and some parts of Lahoul-Spiti (Himachal Pradesh). Truly described as cold desert with low population density, Ladakh constitutes the easternmost trans- Himalayan part of Jammu and Kashmir State of India comprising of Leh and Kargil districts, bordering Pakistan and China. Leh covers an area of 82665 Km2 and 51358 ha of reporting area, situated along the valleys of the Indus River with an estimated population of 117232. Leh is India's highest district, and one of its most arid, coldest and sparsely populated parts (Humbert-Droz, 2004). The district is sandwiched between two Himalayan ranges, Zanskar in west and Ladakh in east. Across the eastern range are the great highlands of Leh district, Chang Thang, which extends along the south of Ladakh and west, over the Indian-Chinese border into Tibet. To the north of Leh district is the district of Kargil and beyond it is the Karakoram Range. The Indus River runs through Leh district from south to north, collecting the water of melting glaciers and straddling several high-lying villages along its banks. Geo-ecologically Leh district is divided into almost two halves- a cold arid desert with minimal precipitation in the northern half, and high altitude grassland (Chang Thang) with perennial grasses in the southern half. The cold desert region of Ladakh, lying at an elevation of 2600 to 5030 m above MSL, is characterized by low relative humidity (20-40%), low atmospheric pressure (493 mm Hg), low partial pressure of oxygen, high wind velocity (8-10 km hr-1), very low annual precipitation (80-300 mm), and subzero temperature (up to -40°C) during winter months. Intense sunlight, high evaporation rate, strong winds and fluctuating temperature (30 to -40°C) characterize the general climate. It is generally said that a man sitting in the sun with his feet in the shade can have sunstroke and frostbite at the same time. With sparse vegetation there is little moisture in the atmosphere. Rains are very rare, though it may even snow during July-August, the hottest months. Because of high mountains all round, the area remains landlocked to the outside world for nearly six months in a year.

Irrigation Technology

Despite the mountains being global storehouse of fresh water (in the form of glaciers), the communities residing in them have to struggle with a severe shortage of water, both potable, as well as for irrigation. Nowhere does this cause greater hardship for habitation and livelihood than in the cold desert regions of the Himalayas. The reasons vary from climatic to geographic to human-induced ones. Depleting water resources are manifest in the form of thinning/ retreating glaciers, sedimentation of water harvesting structures, etc. In the Trans-Himalayan region, the main form of precipitation is snow, but barely 250 mm or less per year. The economy of these regions is based primarily on agriculture, but the cultivation period is very short as the land is snow-covered for about 6 months. Water for irrigation comes almost from glacial melt, transported over long distances through small channels that are locally called canals/*kuhls*. In view of the prevailing arid conditions in the district, agriculture is possible only where water for irrigation is available; much of irrigation in the district is through canals/khuls.

Irrigation technology in Ladakh was transferred from neighbouring regions (Osmaston, 1985). The farmers have partly transformed the barren semi-desert into green croplands through skilful irrigated cropping that goes back to at least Tenth Century A.D., when it is said to have been introduced by the scholar saint, Atisa (Bell, 1928). The major source of water in Ladakh and other similarly cold arid areas is the glacial melt water through streams.

Other sources are the natural springs, which are generally the outlet point of various watersheds, and whose discharge is controlled by groundwater from various glacial-melt streams. More often the streams run some distance away from the cultivable land, or are incised deeply below it, so that a long canal is necessary to bring water to the desired place.

It is not easy to lay the canal avoiding the natural obstructions, or to maintain a uniform gradient. It requires considerable traditional expertise. The skillfully engineered work sometimes gives rise to an optical illusion that it is flowing up-hill. The canal bed is often made of very porous material, loose stones and boulders, so there is considerable loss from seepage. The bushes and trees growing around the channels confirm that these losses take place. The Stongde Gompa (in Zangskar valley of Ladakh) canal loses nine/tenths of the initial flow of 0.01 m3/sec before reaching the Gompa (Crook and Osmaston, 1994). Wherever it is possible, farmers have skilfully diverted the water through construction of long canals, some of them running for few kilometres traversing through rocky mountains.

These canals were constructed during early stage of the history of the region. The glacial melt waters form various rivulets called *kangs-chhu* (ice water) which join together to form a *togpo* (stream). Togpo water is discharged/diverted to a valley touching many villages, in which the main canal, called *ma-yur* (mother channel), is constructed. It is built along a mountainside that forms its retaining wall, and is lined with clay to hold the water. This may be termed the Ladakhi version of a dyke. Water from the *ma-yur* is further diverted into *yu-ra* (small canal), which irrigates the fields. The point from where togpo water is diverted into *ma-yur*, and *ma-yur* water into a *yu-ra* is called yurgo. Ska is the point from where *yu-ra* water is diverted to the field. Water in the ska is further guided through channels known as *snang*, which carry the water throughout the field. The cultivated fields are generally on terraces and irrigation is through gravity gradient. Therefore, immense skill is required of the farmer to maintain the gradient of the field. The gradient should be such that it is neither steep nor flat, rather a gradient that allows smooth flow of water with enough seepage into the soil with negligible erosion. The water distribution through a system of channels is quite complex with different sizes of channels distinguished by various names. There may be some regional differences in the phonetics of the names. Moreover, it is similar in whole of Ladakh (Crook and Osmaston, 1994).

Ponds

In some villages, water from the *togpo* through *ma-yur* is stored in rdzing (pond). The water from this is then diverted into a *yu-ra* that carries it to the field. Individual families rarely construct their own pond. Every year in the beginning of spit (spring), ponds are cleared of silts. The villagers collectively undertake the cleaning operation. Usually the silts are piled up into *lut-pung* (small mounds), on the edge of the pond to hold the water, as this silt is used to prepare manures it is stored in dry lavatories. Usually a pond gets water supply from the village stream through the *ma-yur*. In some villages like Phey, a-Yu and Ska-ra ponds are also supplied water from *chhumig* (spring water). This is possible because they are situated at a lower level, and because the springs in these villages are well endowed.

Irrigation Water Management

Irrigation water is a critical source of food and wealth in any high altitude cold desert. One might expect to find a complex politics of distribution and exploitation. Distribution is certainly complicated, but exploitation is rare (Gutschow, 1993). There are inevitable abuses and conflicts in the distribution of water. Villagers are continually being watched at the same time that they are watching out for their neighbours, thus creating a reciprocal check on activities. It is difficult to evade signs of cheating, a moist field when all the surrounding fields are dry. The combination of an effective social organization with technical expertise suited to the local environment has enabled the farmers of cold desert to convert a few patches of their semi-desert to a very intensive and highly productive agriculture. This adaptation by the Ladakh farmers to a high altitude cold desert is the foundation of human settlement in such an inhospitable environment. The farmers in large villages are divided into groups commonly known as *Schhu-cho,* and each group gets right to water according to the traditional distribution system. The groups and the distribution pattern of water are recorded in the land records, stored at the patwari's office, whose duty includes maintenance of an official register of these water rights or on a silk document in the village itself, generally referred to as *bandabas* by local people. The documentation suggests that these disputes could be taken to the court. While such cases are beginning to be witnessed in some rare cases, mostly villagers still prefer to negotiate their disputes at the village level itself. Distribution of water for irrigation is in accordance with the rotational system. The rotational system is largely determined by the village topography, total village acreage, relative exposure to sun, average temperature, size of glacier, soil type and seepage in the irrigation channel, among other factors (Gutschow, 1993). The amount of acreage and the number of irrigation channels in any given village seems to determine which kind of rotational scheme is used. If the village is small and the irrigation channels few, rotational system of water distribution is pegged to the household, rather than the channels. For example in Leh and Sakti, water distribution is arranged by channels. In Phey (small village of forty-three household), water distribution is pegged to the household. This is because in small villages it is fairly easy to arrange and anticipate whose field will get water according to the rotational system. In a village where distribution is arranged by channels, the field lying along the given channel is irrigated in order, which may belong to a large variety of households. These predetermined rotational schemes between households, are in accordance to *bandabas*.

Chhur-pon

The word *chhur-pon* means Lord of the water, derived from *chhu* meaning water and *spon* meaning Lord. Water supply to individual families for irrigation is supervised by a *chhur-pon*. The *chhur-pon* is an official selected by the villagers, who is in charge of water distribution for irrigation and is perhaps the most important functionary in this regard (Koshal, 2001). This functionary used to be selected by consensus in earlier times. Nowadays rotation system is prevalent in most villages. The office of the *chhur-pon* has been in existence ever since people can remember, except in villages where water is available in abundance. In some villages the word *chhur-pon* is also assigned as individual house name (in Ladakh every house has a name). This is a clear indication of the importance of *chhur-pon* in these villages. If water is available in abundance then a *chhur-pon* is not required, as in the case of Hunder village of Nubra block. Even in villages where there are *chhur-pons*, his task is reduced to a great extent when water is in abundance. If the scarcity of water in a village is acute, then more than one *chhur-pon* is required, even if the village is small in size. In return for their service a *chhur-pon* is given *so-nyom* (one man load of cereal crop) after the harvest but nowadays this has been taken over in monetary terms, and the amount differs among villages. The *chhur-pon* has thus quite difficult duties to perform. As a customary respect for water, while stopping or releasing water from *ma-yur* to the *yu-ra* the *chhur-pon* has to keep his *gon-cha* (long overcoat) at ankle length and can not tuck it into his belt as is often done while working.

It is expected of the *chhur-pon* to distribute water according to the rota system, and monitor the activities of other farmers. At the operational level he is not much seen in the picture. Members of the village monitor each other's activities, and this system is inbuilt, as all the members are aware of how much and at what time water has to be diverted into their *yu-ra*. The farmer receiving water first monitors that the water is not diverted earlier than the decided time, and the farmer who receives later, monitors that his turn for water is not delayed. The first watering of the field occurs soon after the *sa-ka* ceremony is performed. Sa-ka is a festive-auspicious occasion. The day for *sa-ka* is decided by the village *onpo* (astrologer). On this day, all the work concerning cultivation, including construction of *ma-yur, yu-ra,* their repair work, etc., as well as the order in which it is to be done, is decided, and appropriate persons from the village are selected for various jobs. Only after making all the arrangements the agricultural work at the village level gets started. The *sa-ka* is performed in each village, or sometimes a group of villages performs it jointly. Its time varies from valley to valley or even from one sub-valley to another. It was customary to celebrate *sa-ka* ceremony on *gya-pe ma-zying2* (the mother field of the king), who possessed land in every village. After the conquest of Ladakh by the general Zoravar Singh, *sa-ka* was held in the *mazying* (mother field) of the aristocratic family. In villages where there is no aristocratic mansion, it is performed in the main field of the monastery. Besides, each family performs *sa-ka* on their own *ma-zying.*

The amount of winter snow determines the anticipated supply of water during summer, which further determines whether marginal fields should be sown or not in spring. The weather in spring determines whether the irrigation water supply will start early or late, since cool cloudy weather delays the snowmelt. This affects the ploughing and sowing time and the sequence in which the crops are sown. The application of sufficient but not excessive irrigation water is one of the most important farmer's skills. Insufficient water will result in poor growth, reduction in yield and possible risk of salinization. Excess can be dangerous to newly germinated crop, leach nutrients and waste of water. To avoid this situation when the crops are tender, only the skilled persons, especially the elders, are involved in irrigation of the field (mostly the first two irrigations, after the seeds are sowed). Later, when the crops are established well in the field, the less experienced younger people provide irrigation. Fields are irrigated for the last time in autumn after the soil is ploughed when harvesting of crop has been done. This practice causes the water to freeze in the soil, and prevents it from being flow out. The water is available again when the spring thaw occurs, and the soil gets moist for ploughing.

Watering a field

First Watering (tha-chus): Once the field is irrigated it is left for 3-15 days depending on the quality of the soil. The moist earth is called ser (gold), more precisely *chhu-ser* (literally *chhu* means water) i.e. earth is gold when it is supplied with right quantity of water. Soil with the right moisture is best suited for sowing and is referred to as *ser- phar- tog* (gold is ready); if it is too damp, then it is *ser-lchin-te dug* (gold is heavy). A snowfall in winter is a source of great satisfaction, for when it melts the earth absorbs the water and remains moist till ploughing and sowing begins. This kind of moist soil is called *kha-ser* (earth made gold by the snow). This saves farmers from first watering of the field and also the drudgery of waiting for their turn.

Second Watering (dol-chhu): After the field has been ploughed and sown, the second irrigation is given. Around 15-20 days of gap exist between the first and the second irrigation. The second irrigation is most important and is a delicate task because by this time the seeds have just sprouted: *dol* (to sprout), *chhu* (water), i.e. watering the tiny budding shoots. A skilled and knowledgeable person takes the responsibility of irrigating the field at this stage. Usually second irrigation needs more than one person. Insufficient or excess watering may lead the tender plants to get *tshik-ches* (burnt) or *shiches* (die). The *khyem* (spade) used for irrigation also serves as a measuring rod.

Third watering (sak-chhu): It is given 15-20 days after the second irrigation. It is generally a light irrigation, otherwise the crop will die. By this time the crop grows up to a height of about five inches (~13 cm). If denied irrigation at the right time, the upper portion of the crop dries up. In case someone's field is in great need of water, but his turn to irrigate is far away, then he requests someone irrigating the field to release one *sha-gugang- ngi* or *nang gang-ngi chhu* (a channel-full water).

Fourth Watering (non-chhu): It is given 10-15 days after the third watering. By this time the crop grows strong enough to stand a little excess or slightly inadequate supply of water. Anyone can look after the watering because it does not entail any specific knowledge.

At the time of *non* (all the crop has grown to the same level) *chhu* (water) it is easier to find out which plants have withered away and which would survive. In other words, the condition of the field becomes clear. From now onwards the crop has to be watered regularly, about once a week depending on the moisture in the soil and weather condition. No special care is required for these watering except that the water is let in and then closed. These are generally undertaken by the less experienced younger people. These continue for about six times before the last watering, which is called *do-chhu.*

Last Watering (do-chhu): Application of *do-chhu* is very important. If the field is not watered now the ears of the crop get dry and start falling. *Dochhu* means watering of fully-grown crop field. The shoots are of equal height and ready to be reaped. By this time some members of the family water the crop for the last time, while other members are busy harvesting the *ol* (alfalfa).

Global Warming: Ladakh Perspective

The evidence of global warming is becoming abundantly clear with newly appearing reports on climate change. One of the major consequences of rising global temperature is that the glaciers and ice-shelves are receding. This is true for the cold desert of Ladakh also (Jogesh, 2008). With diminishing glacial resources and increasing variations in climatic and ecological conditions, impacts like flash floods and frequent locust attacks are increasing. The Leh's inhabitants are, therefore, trying to adjust to the new threats to their traditional livelihoods and sustenance. The Tsomoriri lake in the eastern Changthang valley is glacier fed so anything that happens to the glaciers will impact the lake. The locals say that there are signs to suggest that the water level of the lake has increased over the last couple of years. Trekking along the Tsomoriri wetlands to a small island in the lake is a place that was once a breeding ground for the highly endangered black-necked crane. Today, with rising water levels, the island is almost fully under water and the last crane was spotted here seven years ago. Only 60 black necked cranes exist in India and experts say, that if global temperature continues to rise, more than 100 million species will be at the risk of extinction. Ladakh appears to be at the starting point of the climate crisis in the region, which if continues, will threaten not just the biodiversity of the whole region but also most of its water resources (Jogesh, 2008). According to Behera and Rahul (2008) ecologically fragile geographical areas like Ladakh are most likely to be adversely impacted by climate changes, which can rightly be considered as global commons, often inhabited by traditional communities dependent on scarce natural resources for their livelihoods, which are often at subsistence or near-subsistence level. Hence, an understanding of these communities' perceptions and responses to climate change are essential to constructing effective sustainable development strategies that mitigate adverse impacts of climate change. They found that the perceptions of climate change are widespread and include observations of rise in temperature and heavy snow melt, less snow fall, heavy and untimely rainfall, biodiversity loss, etc. These changes have severely affected the use and management of natural resources and hence the livelihoods of the local people. Leh district receives about 20 mm rainfall in a year, which has risen to nearly ten times in the past few years (Sethi, 2007). Evident more than 73% of the surveyed households in Leh district have perceived the increasing intensity of rainfall. This heavy and untimely rain in recent years has caused widespread damage to private and public property through landslides and flash floods. The flash flood of August 2006 in Leh destroyed homes, bridges and roads, cutting off villages from sources of communication and assistance, and inundated the fertile fields with rock debris and sand, rendering the land unsuitable for further cultivation. Inhabitants of Leh district feel that climatic change has affected their livelihood in some way. According to surveyed households despite a well functioning *Chhurpon* system, their villages face severe water scarcity, which is attributed to heavy runoff of water from faster snow melt, and erratic nature of sunshine hours causing unpredictable flow of glacial melt water into the streams. It is important to note that the run off water enters the Indus River and flows downstream. Ladakhis have traditionally not relied on the Indus for irrigation (except for few villages like Shey, Thiksey, Choglamsar, etc.) due to its shallow depth and difficulty in lifting water from its levels without the use of less efficient mechanized pumps that also pollute the environment. The altitude and climate of Ladakh pose major mechanical challenges that are often not experienced in any permanently inhabited part of the world (Behera *et al.*, 2008).

The disappearance of snow from the surrounding mountains has led to reduced grass yields important for grazing. As a result, the pastoralist groups now move more frequently from one place to other in search of grazing lands as compared to a decade earlier. This has further impacted the regenerative ability of the highlands. Another impact of climatic changes has been the increase in the frequency and intensity of pest attacks. The recent (2006) devastating invasion by locusts, both in magnitude and duration in Nyoma and Durbuk blocks is perceived by inhabitants of villages such as Anlay, Rongo, and Kuyul in these blocks as a consequence of the rise in temperatures in the region. It has been noted by the agricultural extension office and the Defence Institute of High Altitude Research (DRDO) at Leh that the frequency of such pest attacks can be attributed to warmer temperatures which is conducive to the increase in the geographical range and the duration of infestation of such agricultural pests.

2

Forestry and Agro-forestry Management Practices

Silviculture

Wood Lots

i) Sacred wood lots: It is an old traditional practice to conserve and protect an entire forest or a small wood lot in the name of the village Deity (*Kul Daveta*). This is prevalent in every village of the mountainous region of India, particularly so in Himachal Pradesh, Garhwal and Kumaon hills and also Kashmir hills. A temple along with a hut or shelter is constructed to keep the items/utensils of the village Diety. All religious rites are performed in its complex and no body is allowed to fell the trees. This is also a common practice in the hills of Kangra, Mandi and Shimla districts. It is also seen that a perennial water source is managed either in the forest or near it, for common use by the village for drinking purpose.

ii) Private and community wood lots: Community and private forests (wood lots) are protected at the community level. In community forests, the whole area is divided into small block or compartments and each compartment is allotted to the inhabitants for rotational lopping of fodder particularly oaks woodlots during lean (winter) season. Frequent fuel wood collection is permitted only in case of twigs and/or branches which are carried and stocked in a traditional manner.

Villagers employ or engage a caretaker for the protection of private or community forests. No one is allowed to lop and cut trees for fuel wood. If some one is found doing so, the caretaker confiscates his/her sickle/axe and ask him to appear before the village panchayat, which imposes a penalty on him in the form of monetary fine or otherwise. Every household in the village contributes some grain or cash in each crop season for the caretaker's maintenance. In water logged areas along with *kuhal* or *nallas* people raise plantations of *Salix* (Bhains), *Alnus* (Kunish) or other moisture loving trees. Farmers make extensive use of the wood from both private and community forests for building all kinds of houses, shelters etc. and have devised their own technologies for postharvest processing of the wood.

Katha Extraction

The heart wood chips of *Acacia catechu* trees are boiled in earthen pots along with water for about 12 hours or till the boiling mixture attains a thick syrup like consistency. This thick liquid is poured into moulds dugout in fine sand which absorbs the tannic acid (liquid) and leaves behind a residue of catchin (Katha) in a crystallized form. On cooling it solidifies and then is cut into small bricks for sale in the market under the trade name of 'katha'. In India it is consumed with beetle leaves and used medicinally for intestinal disorders and is known to improve digestion.

Traditional Minor Forest Resources Management

In the lower hills of Himachal Pradesh, Uttar Pradesh and of Jammu & Kashmir, *Acacia catechu* (Khair) for katha, *Pinus roxburghii* (chirpine) for resin. *Acacia nilotica* (Kikar) for tannin and gums and *Grewia optiva* (Beul) for fodder and fibre are well known. The villagers manage and protect these trees/wood lots and obtain high economic returns. These wood lots are inaccessible for fuel and timber requirements. The villagers also get seasonal employment in these forests for the collection of value added minor forest products. Similarly in the Kinnaur district of Himachal state, which is known for the production of *Pinus gerardiana* (chilgoza) seeds and *Prunus armenica* (chuli) fruits, the local people have right to collect the same from the forest and to sell them in the market for economic remuneration. The Government forests of these species are managed by the villagers as common properties. These forests are divided into small blocks and each block is allowed to a family for protection, management and collection of produce.

Agri-silviculture

Jhuming or Shifting Cultivation

It is a widely practiced farming system in the hills of north-east India, locally known as "podu" cultivation. It is believed that this farming system originated in the Neolithic period around 7000 BC.

This system involves cultivation of crops on steep slopes. Land is cleared by cutting forests, bushes etc. upto stump level in December-January, leaving the cut material for drying and then finally burning it to make the land ready for dibbling of seeds of different crops before the onset of rains. The cultivation is confined to the village boundary and the total area brought under cultivation depends upon the family size. After 2-3 years, the area is abandoned for regrowth and a new site is selected to repeat the process.

Shifting cultivation was good at the time when it emerged, because mixed cropping on slopes under rainfed conditions and a dependence on local natural resources only kept disturbance of the soil to a minimum and furthermore its fertility was allowed to replenish adequately. But now, due to increase in both the population and the effective area under this farming system, it is considered to be hazardous because it is leading to resource degradation and ecological imbalance due to deforestation, increase of soil erosion and silting of reservoirs which further lead to floods. It is also a low technology and high labour intensive system.

Various research institutes of ICAR, GOI and State Government working in the north-eastern part are recommending at scientifically tempered Agro-horti-silvi pastoral system as an alternative to shifting cultivation. Despite its adverse effects on the eco-system and low productivity it still persists because it is an integral part of the socio-cultural fabric of tribal life in the northeast and is clearly linked with their religious rites and festivals e.g. Agalmaka, Miamua, Rongchugala and Ahia of Garo Hills.

Homestead Agroforestry System

A homestead or homegarden is an operational farm unit in which a number of tree species are raised along with livestock, poultry and/or fish; mainly for the purpose of satisfying the farmers' basic needs. This farming is traditional in the Himalayan region. In a home garden multiple crops are present in a multi-tier canopy configuration. The leaf canopies of the components are arranged in such a way that they occupy different vertical layers with the tallest components having foliage tolerant to strong light and high evaporation demand and the shorter components having foliage requiring or tolerating shade and high humidity.

In Hamirpur and Kangra districts of Himachal Pradesh the homestead is constituted of Fodder trees i.e. *Celtis australis* (Khirak), *Bahaunia variegata* (Kachnar), *Grewia optiva* (Beul) in *the* upper story. The middle storey is constituted of bushes like *Adatoda vasica, Vitex negundo,* lemon and Galgal as fruit trees. During the rainy season cucurbits (vines) are grown along with Colocacia, turmeric in the field and (winter) vegetables during the winter season.

In Meghalayan homestead pisciculture is the common practice. Fishes are generally reared in a dugout ponds either in a comer or in the centre of the home garden, so that run off can be collected in it. In the low lying or plain parts paddy is generally grown. Important timber trees are *Pinus kesia* (Khasipine) *Alnus nepalensis* (Utis) *Schima wallichii* (chillowane) etc. are grown in the homestead for their requirement.

In the lower storey guava, banana and vegetables are grown for domestic consumption. As per the study conducted by the ICAR Research Complex for NEH Region, Shillong; in a dugout pond of 32m × 15 m 2.2 m size there was a production of 250 kg fishes in one season during the year 1990.

However, wide variations in the intensity of tree cropping are found among homegardens in different locations. This can generally be attributed to the differences in socio-economic conditions of the households and their responses to externally determined changes particularly to price of the inputs and products, dependence on land and tenurial conditions among others.

Plantation Crop Combination

Plantation crops play a major role in national economics because they generate value added goods for the international market. The important plantation crops of the Himalayan region is tea. Traditionally, tea is grown on waste and marginal lands either in monoculture or in association with indigenous forest tree species.

In Himachal Pradesh tea gardens in Kangra Valley are managed under the canopy *of Albizia chinensis* spp. which not only nurse the tea plants by fixing the atmospheric nitrogen in its roots but also provides shade for the development and maintenance of new tenders. The leaf litters *of Albizia* trees also add nutrients to the soil. Similarly in Assam and West Bengal the tree gardens are managed under *Alnus nepalensis* (Utis).

Large Cardamom Based Agroforestry System

Large cardamom (*Amomum subulatum*) a traditional plantation crop is grown in an old typical agroforestry system. It is adapted to humid and shady environment and requires moist soil with good organic content. In natural forests it is raised in association with *Alnus nepalensis* (Utis), a shade tree adjoining to small perennial streams (Thoras).

This helps to spread water seepage through the soil during most part of the year. *Alnus nepalensis,* fast growing tree, not only produces fuel wood and timber but also fixes atmosphere nitrogen and enriches the soil by adding litter which in turn benefits the large cardamom. It also serves as a catch crop; attracting leaf feeding insect-pests to itself thereby protecting the large cardamom crop.

This silvi-horticultural system is very successfully practiced in Sikkim since over a century. The area reported under large cardamom is about 14, 000 ha in Sikkim and the acreage is expected to increase phenomenally with the aid and assistance of the Cardamom Board established in Sikkim recently. In Joginder Nagar, distt. Mandi (*HP*), cultivation of large cardamom is picking up under *Celtis australis, Ficus* spp., *Alnus* spp. etc. near the perennial streams or *nalahs.*

Bamboo Groves as a Component of Agriculture Holdings (Agri-silviculture)

In Sikkim agriculture is well established as compared to other northeastern hill states. The farmer in Sikkim, by and large, adopts a mixed farming system. In all agricultural holdings along the streams and irrigation/drainage channels cultivation of bamboo (*Dendrocalamus hamiltonii* and D. *sikkimensis is* common place. Bamboo's are extensively used in building small farm houses, goat sheds, piggery enclosures, small baskets and string making. Large bamboo pipes are used as water conveyers in the farm irrigation/ drainage system. Bamboo leaves serve as an excellent winter fodder for goat. Bamboo stumps/culms also protect the water channel from erosion. This system is limited only to he high rainfall areas or where the sufficient water is available to grow bamboo. A similar practice is also found in the Palampur area of Himachal Pradesh as well as in other northeastern states.

Multipurpose Trees on the Crop Land

It is a common practice in the western Himalayan region to cultivate wheat, peas, potato, cauliflower, mustard etc. during winter and maize, tomato, chillies etc. during the summer season either in monoculture or mixed on the permanent terraces prepared across the hill slopes. Fodder, fuel and timber trees viz. *Grewia optiva* (buel), *Celtis australis* (Khirak), *Bauhinia variegata* (Kachnar), *Albizzia chinensis* (Ohi), *Toona ciliata* (Toon), *Morus alba* (Toot), *Ulmus laviegata* (meryano), etc. are deliberately left or grown on the bunds of terraces. This type of practice i.e. agrisivicultural system is prevalent in sub-montane and mid hills sub-humid zone of Himachal Pradesh.

Similarly in the Agri-horticultural system which is predominantly followed in the mid hill-sub humid and high hill-temperate wet zones fruit trees are grown on terrace bunds along with agriculture crops since fuel and fodder are readily available from other sources. Agricultural crops like peas, cabbage, colocacia, turmeric and pulses are generally grown in the inter-spaces of horticultural trees such as *Malus domestica* (apple) *Prunus domestica* (plum), P. *armeniaca* (apricot) P. *persica* (peach), P. *deueis* (almond) and *Pyrus communis* (pear) Tree are uniformly spaced in the field and average density of trees is generally 5 per 100 m^2. In some cases fuel or fodder/timber trees are also retained on the field bunds and managed under agri-horti-silvicultural system. Besides providing fruits/fuelwood/fodder trees grown along bunds also restrict/.check erosion.

Private and Community Fodder Wood Blocks

Fodder tree forests in the vicinity of villages, be they on forest land, community land or private land are jointly managed by the managers to meet their fodder need for cattle during lean period i.e. winter season. In the mid and high hills of Himalayas *Quercus* spps. (Oaks) forests are jointly managed and protected for the same whereas in the foot hills or sub-mountainous region other broad leaved trees i.e. *Grewia optiva* (Buel), *Celtis australis* (Khirak), *Bauhinia variegata* (Kachnar), *Albizia chinensis* (Oei) etc. are managed so as to meet fodder and fuel requirements.

Seabuckthorn: Potential Resource

Seabuckthorn (*Hippophae* L.) an ecologically viable and ethnobotanically sustainable food crop, belonging to family Elaegnaceae, is one of the few potential resources of high mountain areas which has been reported to offer hope to give multiple benefits to poor mountain man. Seabuckthorn, (an indigenous plant which has survived under harsh climatic conditions such as average rainfall 350-1200 mm, average temperature in the range of 10-15°C and sometimes going below -40° to -50°C), is one such specie which has immense potential for satisfying all requirements. Seabuckthorn is widely distributed along high hills, river beds, valleys and along dry mountain slopes of this region both as healthy dense stands as well as in scattered gregarious dense patches in an altitudinal range of 700 in to as high as 5500 m and above.

The people in high mountainous areas have been using seabuckthorn fruits for curing several ailments, including skin tumor diseases, digestive disorders and respiratory ailments, foot and mouth disease in cattle, besides using it extensively in agroforestry practices, though the farming community did not know of its exact potential. Seabuckthorn has also been a major source of fuelwood to local inhabitants who face scarcity of other sources of energy. Major traditional uses of Seabuckthorn in high mountainous cold and dry zone of Himachal Pradesh are fuelwood, fodder for goats and cattle (nutritious forage), fencing to protect fruit trees/nurseries/orchards, maintenance of traditional irrigation channel besides soil conservation, improving fertility/quality timber, useful for newly opened areas and subsidiary food items, medicinal products and alcoholic preparations.

In Himachal Pradesh major gene pool areas have been identified at Spiti valley, Lahaul and Kinnaur, for 4 species and 9 subspecies of genus *Hippophae* reported from different parts of the world.

The introduction, promotion and development of seabuckthorn in this area seems meaningful and will not only help in vegetation rehabilitation, but also in ecological sustenance and economic gains to local farmers of the cold and dry mountain area.

Plant Protection Practices

In recent years, the large scale application of pesticides, primarily insecticides, has made cultivation of some mountain crops and fruits difficult due to the harm brought to bear upon the bio-environment by large scale destruction of natural bio-enemies causing pest resurgence, development of resistance to pesticides and consequent secondary pest outbreaks. Perhaps the only solution to this problem at hand lies in the adoption of eco-friendly approaches which are not destructive to natural enemies but gradually remove sizeable proportions of pest populations and tend to keep their populations in check. Physical, (devices and procedures used to change physical environment of pest populations), and mechanical (mitigating pest populations by cultural practices) methods of pest control are the oldest of all such insect control methods. These are rooted in simple practices that man, as a farmer, has leant from his long and close association with pests. These aid him in reducing pest populations to low levels. These include both direct and/or indirect measures which may be preventive or corrective in nature but are essentially slow acting, often ecofriendly, cost effective and compatible with other methods of pest control. These characteristics make them amenable to blend better with integrated pest management practices (IPM) even though they do not bring about an immediate or drastic reduction in pest populations. Even the modem concept of pest control does not emphasize the outright eradication of pests but focuses on maintaining their populations at levels which do not cause economic losses. Some of the indigenous methods of IPM include:

Ploughing, hoeing and basin preparation

Cultural practices like ploughing, hoeing and basin preparation influence directly, the survival of soil inhabiting pests. These routine agricultural operations expose soil inhabiting insect, pests and other arthropods and nematodes to harsh weather and to natural predators. Insects are most vulnerable when in the pupal stage and most insect-pests pupate in the soil which furnishes a protective habitat. Birds like the king crow, the myna, the starling, etc. pick up the exposed pupae following these cultural operations. Some insects e.g. grasshoppers, crickets, mole-crickets and borers lay then-eggs in the upper layers of the soil. Their eggs are exposed during soil preparation and subsequently desiccate. Many insects like cutworms, grubs of the root borer and white grubs which feed on the root system of plants are also exposed to the vagaries of the elements during basin preparation and hoeing. Deep ploughing carried out during winter helps in reducing the overwintering populations of several pests.

Beside dislodging the pests from their protective habitat and subjecting them to unfavorable conditions for survival, these scientifically tempered cultural practices also improve aeration of the soil and facilitate proper percolation of water into the soil.

However, the degree of success of these operations is related directly to the presence of natural predators in adequate numbers and the synchronization of these operations with the vulnerable stages of the pest's life cycle.

Hand Picking of Pests

Hand picking of pests and their destruction is another time tested method of pest control. Right from picking lice from human hair, clothes and even animals to the manual separation of pests from stored grain.

This method can prove effective in curtailing pestilence on some crops. Insects which lay eggs in conspicuous and easily eliminated masses, e.g. tobacco caterpillar, gypsymoth, hairy caterpillar, sugarcane top borer, epilachna beetle, etc. can be easily eliminated. Early instar larvae of such insects often feed in congregations and later disperse on to the entire plant/tree and in the field, making control, at best, difficult.

Hand picking demands alertness, patience and keen observation. The collected pests are destroyed by immersing them in kerosinized water or by deep burying. Nocturnal insects responding positively to light, e.g. defoliating beetles, moths of Bihar, hairy caterpillar, tomato fruit borer, tobacco caterpillar, and cerambycid beetles etc. are collected, using light source or by trapping them in a light-trap and are subsequently destroyed.

However, it is a labour intensive and time consuming process. Knowledge of egg laying behaviour, location of eggs and larvae on infested crop is essential Hence, only a well trained person would prove effective.

Cow-dung and Clay Mixture

The practice of applying a thin paste made of cowdung, clay and cow urine, as a disinfectant, on the floor of mud (*kutcha*) houses in rural areas is an age old practice. Similarly, the application of a paste, of good consistency, to treat wounds and injured limbs of fruit trees has been in vogue, in villages, since long. Such a paste is also applied to pruned twigs/limbs.

Fine slurry is prepared by thorough mixing of the clay, cowdung and cow urine in a container. The paste is then applied manually using bare hands or a locally made brush.

Cowdung and cow urine possess complex degrading substances and may possess anti bacterial properties. Addition of clay results in better adhesion of other constituents to the treated surface. Sealing of wounds/cut prevents access of pathogens to the otherwise exposed surface. To some extent, the already present pathogen is also dealt by using the applied paste. Pruned ends of twigs and cuts are also favourable spots for the settlement and establishment of woolly (apple) aphid on apple trees. Covering such sites with the cowdung paste hastens healing and prevents aphid settlement.

The paste is also used for the treatment of cankered limbs of the trees. The entire effected region is removed by a disinfected knife and is then covered with freshly prepared cowdung paste. As is evident, the application of the paste is a laborious procedure and proper sealing of the injured region is required. Sometimes, it does not prove to be an effective method for the treatment of canker.

Pruning of Fruit Trees

Pruning and training of fruit trees is an important practice performed during the dormant season to create a proper frame and provide symmetry to the tree and ensure proper and balanced growth in the ensuing season. Set procedures of pruning are followed by using secateurs, knives, pruning saw, etc.

Some inconspicuous pests which hibernate as egg nymphs (e.g. some aphids, leaf hoppers, etc.) in limited numbers are automatically removed along with pruned branches/twigs thereby mitigating the severity of their attack in the next season. Further, pruning of old, damaged and weak portion of the tree encourages new growth which is healthier and stronger.

Thus, pruning aids pest control as it realeases overwintering population, improves general health of the tree and helps in maintaining a balanced foliage distribution with proper aeriation and sunlight penetration. Pruned plants usually have lesser pest infestation and do not easily succumb to pest attacks.

However, since pruning is both an art and a skill it demands that the pruner posses basic knowledge of overwintering behaviour and sites of pests to deliver maximum benefit from this mechanical method of pest control.

Use of Wood Ash on And Around Vegetable Crops

Fruit wood is a major source of energy in the hills, consequently, wood ash is available in plenty. It is used to wash and clean utensils as well as clothes in many areas of Himachal Pradesh. It is a common practice to sprinkle wood ash on vegetable crops, especially growing in kitchen garden and to spread it around plants to ward off pests and to enhance nutrient status of the soil. To achieve this a thick layer of ash is spread on the soil around plants and it is also sprinkled on foliage to protect it against a variety of pests.

This is because it is a source of phosphorus for plants and it also acts as a physical poison usually causing abrasion of epicuticular waxes and thus exposing pests to death through desiccation. It also interferes in the chemical signals emanating from the host plants thus obstructing the initial host location by pests. The treated foliage further becomes unpalatable for foliage feeders like cutworms, caterpillars, grasshoppers, etc. But since ash provides only temporary protection against pests, insecticides which have quick knock down effect have replaced the use of wood ash today.

Beating Drums and Using Domestic Dogs for Combating the Menace of Birds and Monkeys

Birds and monkeys are commonly found vertebrate pests which cause tremendous damage to some ripening crops and fruits. Most bird species are protected by law and monkeys can also not be killed due to religious sentiments. Alternately, the practice of driving them away by beating drums and using well trained dogs has been used since long. Locust-swarms are also warded off by the beating of drums. Today, one or two persons are engaged in these activities during periods of fruiting/crop maturity (along with one or two trained dogs *Gaddi* or Alsation). While the noise created by the beating of drums and barking of dogs scares birds and monkeys, it requires the employment of regular labourers for constant vigil (Also monkeys and birds may become use to these practices).

Kerosene Oil for Killing Borers

Shoot and stem borers bore and riddle the twigs, branches and even the main trunk of fruit trees and push out faecal matter and frass through an exit hole. Farmers have been using kerosene oil to kill these control tissue borers using a flexible metallic wire which is inserted through the hole made by the borer into the gallery to clean it. Then a small bung made of cloth soaked in kerosene oil is inserted into the hole and finally, it is plugged using a paste of cowdung and clay. The insertion of the metallic wire into the gallery causes physical injury to the larvae which are destroyed in the process. The oil vapour emanating gradually from the cloth bung fill the closed gallery, suffocating the pests and ultimately the larvae die. The death of the borer is indicated by unopened plugged hole. The drawback of this practice is that kerosene vapours act slowly and this treatment is not a sure and definite method to annihilate the borer. Farmers now prefer to use quick acting insecticides for assured results.

Use of Traditional Devices (Sling and Karban) for Scaring Birds and Stray Animals

In wet temperate Himalayan regions *the* Sling (*Gulel*) and the *Karban* are commonly used for scaring away birds and stray animal for the protection of crops such as maize, fruits and vegetable.

Bajagun Weed for Reducing Rat Menace

Common cowitch, *Mucuna prurita,* vernacularly known as *Bajagun,* is a leguminous herbaceous annual plant found throughout the plains of India. It also grows in the low lying areas of Himachal Pradesh. Its longitudinally ribbed turgid pods are covered with dense pale brown or grey stinging bristles (trichomas) which upon coming in contact with the skin cause discomfort and itching attributed to presence of a histanine liberating proteinase, mucunain.

The effect persists for 3-5 minutes. These hair also contain serotonin which causes cutaneous pain besides itching. It was an old practice to powder leaves and pods of the plant and scatter it over men-ways and in the live burrows of the rats to drive them away from cultivated fields to reduce rat damage to the crop. After prolonged boiling and throwing away of the water, the pods are also reported to be used as food during famine. The easy availability of various effective rodenticides from the market have resulted in this practice gradually dying out.

Use of Walnut and Sweetflag Leaves Against Pests in Stored Grains

In rural areas of Himachal Pradesh, it is an old practice to use walnut leaves and leaves of a pond weed, commonly known as sweetflag, *Acorus calamus* as a protection for both grain and clothes against insect damage. To achieve protection a layer of leaves of walnut is spread over grain stored in gunny bags. Likewise, shade dried leaves of sweetflag are powdered and put over grain stored in gunny bags to protect it from damage due to stored grain pests.

Walnut leaves are astringents and the aqueous extract of fresh leaves possesses bactericidal action while mature leaves contain 9-11 per cent tannin. Tannins are known to act as feeding deterrents. Sweetflag leaves and rhizomes have many chemical ingredients including an essential oil, the oil of *calamus,* which primarily contains *asarone.*

Depending upon quantity of *asarone* (*cis* as well as *trans*), its effect on insects may be attractant, antifeedant, repellent, antigonadal or insecticidal. However, since the active principal is present only in meagre amounts (in the leaves) this treatment may not be able to afford protection for long periods.

Indigenous Beekeeping Practices

Rich indigenous knowledge of beekeeping in a variety of hives such as wall cavities, hollowed logs, skeps etc. served to preserve this heritage. These indigenous hives resemble closely the natural nesting sites of the native honey bee (*Apis cerana*). Swarms of feral populations of this bee descend to colonize these nests. These are manifestations of rural wisdom and some of them are explained here.

The Orchard hives are made of locally available materials viz., stones available on field; clay; chopped wheat straw; *sarkanda* grass; wood for making frames; string and strips of wood.

First, a platform of stones plastered with mud and measuring 70 cm × 70 cm × 50 cm (l × b × h) is made at a suitable site in the orchard/field. It is given a south-easterly aspect. The brood chamber of standard size for *Apis cerana* type hive is made on the platform from stones plastered with mud. An entrance hole of about 1.5 cm diameter is provided in one of the lengths, for the easy movement of bees. The top cover is made from sarkanda grass tied with string and supported with strips of wood. It is fashioned as a slopping roof to drain off water.

The low cost and easy to construct structure ensures better temperature regulation inside the hive, enhances its suitability to temperate regions and is helpful in successful overwintering. Appropriate location in the orchard renders it to be very useful for pollination of the target crop and all the scientific management practices can be taken up. Additionally, the structure affords protection from wild animals and is suitable for stationary beekeeping as it cannot be moved even within the orchard

Straw hives too, are made using locally available materials. These are: rice straw; locally made string for tying straw; strips of waste wood and wood for making frames.

The rice straw fibres are tied into 5 cm thick sheets with the help of strings and wooden strips. These sheets are cut to size and joined to form a brood chamber 310 mm × 320 mm × 204 mm (l × b × h) in accordance with the standard for *Apis cerana* frame hive. A hollow razed of bamboo is placed into one of the lengths to provide entrance to bees. A 5 cm thick crown/inner cover is similarly made from rice straw. The top cover is made in the form of a sloping roof from straw or *sarkanda* grass held in place with sticks. It is left projecting beyond the brood chamber to dram off water. Care has to be taken to ensure that no crevices are left. In addition, no loose straw should project into the brood chamber as it would interfere with bee activity.

This hive is suitable for temperate regions, as its management during peak winter is quite easy and winter-inner or outer packing material is not required. Internal temperature of the hive is also not quickly influenced by changes in environmental temperature. The bees therefore have to spend less energy in temperature regulation within the hive thereby lowering honey consumption and consequently making more honey available for harvesting. Further, overwintering is better as evidenced by larger area under the brood upon the onset of spring.

The wall hive is a cavity left in the wall when the house is under construction; it is located at a height of about 150 cm from the floor in kitchens, store rooms, living rooms and/or sheds; the wall hive has an entrance hole of about 2 cm diameter towards the outside; on the inside it is covered usually with a plant of wood plastered with mud. A swarm of *Apis cerana* descends naturally and settles in this hive making parallel combs. The wall hive is opened only to harvest the honey and never otherwise.

The *Apis cerana* bee's habit of nesting in natural dark and protected cavities makes the wall hive an ideal home for the swarming bees to descend and colonize. The location of the wall hive at a height in farm houses keeps it away from any disturbance due to activities of the farm's habitants. Human and animal habitation of the farm house, however, provides a relatively constant temperature. Since microclimate of the hive interior is not easily influenced by sudden outside changes, the bees have to spend less energy towards temperature regulation. Further the security offered by human habitation adds to the suitability of the wall hive for bees as they are protected against wild animals such as bears and pine martens.

The log hive is a simple structure without any frame and separate chamber. Prepared by the hollowing out of a piece of tree trunk, closing both its ends and boring a small hole of suitable size along its length to serve as an entrance. Both horizontal and vertical log hives are in use. Length of horizontal log hives varies from 60-75 cm and the entrance hole which is 6 mm diameter or of pencil thickness is made in the centre. Open ends are closed with a piece of tin or wooden plank or stone mixed with cowdung and clay. 4-5 strips of old combs (3 cm apart) are fixed at each end before use. The vertical log hive is very similar to the horizontal one except that it is placed in an upright position on a piece of tin or flat stone with 5-6 entrances/ holes above the base. The top closed end may have 8-10 holes of 1 cm diameter each, if the super box is to be used.

Though movable frames can not be used, as such the scientific management of colonies is not possible and only squeezed honey can be extracted. The advantages afforded by this low cost construction makes it amenable to shifting to orchards for pollination and short distance migration can also be taken up.

Post-harvest Management of Food Crops

As Upper Himalayas manifest diverse agroclimatic conditions varying from subtropical, wet temperate to dry temperate, cold desert areas, the different types of agricultural crops grown, vary from cereals, pulses, oil seeds, sugar cane, root crops, vegetables to subtropical fruits viz. mango, *aonla,* citrus fruits, litchi, grapes, etc. temperate fruits like apple, pear, apricot, peach etc. and nut crops like walnut, almond etc. The methods of handling, packaging, storage and preparation of different products and their subsequent consumption also varies with the wide agro-climatic diversity in the region. Most food products currently available in the market are essentially improvements/refinements of indigenous technical knowledge of postharvest management of food crops. However, there are still a number of traditional postharvest skills, which can be commercially exploited. Some traditional foods/products help to cure ailments and have been used as home remedies, while others are environment friendly and do not cause any health hazards despite their continuous use. This is an effort to document available information on postharvest operations of storage, curing and drying.

Curing of Cereals

This is a common practice in the lower elevations of the state and has a scientific basis.

i) The maize plant along with the stalk is harvested and tied into small bundles called *'Pahra'*. All bundles from different fields are collected at one place and stacked in an upright position. Upright stacking of maize bundles is called *'Thua'* or *'Zhumb'*. After a period of one and half month, the maize cobs are separated individually from the husk.

Stacking of maize in the *Zhumb* for 30-45 days, ensures ripening and it also facilitates the easy separation of cobs from the husk. Curing results from heat and moisture equilization in the grains in the *Zhumb*. It has also been observed that initially sweet grains, upon complete drying and curing, become tasteless which is attributable to the conversion of sugars to starch, a common phenomenon also observed in pea and maize grains.

ii) Harvested paddy is tied into small bundles. Each bundle is called *Pooli* or *Poola*. After drying under open condition for 3-4 days, all bundles are stacked at one place and allowed to remain there for few days. Each stack is called a *Kundli*.

iii) Similarly, millet grains called *Mandal* or *Sonk* are allowed to ripen partially by stacking them in a heap and covering the heap with tarpoline. This procedure is locally termed *Garr Dena*.

The preparation of small bundles *Pahra, Pooli or Poola* help in easy handling and quick drying. Stacking bundles in heaps possibly aids heat and moisture regulation thereby resulting in uniform ripening. The creation of moist heat also leads to curing which eases the process of threshing the grains.

Curing of Chilgoza (*Pinus Gerardiana*)

In Kinnaur, chilgoza seeds are separated from the cones by collecting the harvested cones at one place and covering them with chilgoza pine needles, leaves and soil. After 15-20 days, the chilgoza cons are cut-open using a sharp edged axe (*Behla*), with a gentle strike the seeds are easily separated. The procedure of covering harvested cones with pine needles, leaves and soil possibly helps in maintaining desired temperature and humidity inside the heap, which results in the curing of cones.

Cleaning of Grains

i) Winnowing of grain and pulses is a common practice in every home in Himachal. It is performed using a container made of tin, called *Stoop* or *Chhaj*. The grains are placed in the *Chhaj* and slow winnowing leads to separation of dirt and husk from the grain. Almost all types of dry grain like wheat, maize, paddy, pulses etc. can be cleaned in this manner.

ii) Bulk cleaning of grain is done using a container made up of bamboo sticks called *Panaudi*. The dry grain, placed in the *Panaudi*, are allowed to fall from a height of about 4-5 ft in a thin vertical flow in the path of a cross wind. The lighter dirt particles and husk are blown away and the heavier grain is thus separated as it falls straight to the ground. The use of a fan (mechanical or electrical) greatly accelerates this process of cleaning. This method of cleaning is based on the differences in density of the materials to be separated. The use of modern air separators/cyclone separators for grain cleaning is based on this principle.

Packaging of Food Commodities

Fresh Products

Different types of containers are used for packaging food commodities.

In lower areas of Himachal/containers made of bamboo sticks called *Ddalh* are used in the fields for packing grain, maize cobs, potato, ginger, turmeric etc. and also for carrying the material from the field to the house.

For packing clean grain, pulses and flour etc. the *Ddalh* is internally lined with cow dung. The *Ddalh* has a capacity to hold 25-30 kg material. A Similar container, small in size, is called *Tokroo.*

In the higher regions of the state, such containers are called *Kiltas,* however, the shape of this container is different from that of the *Ddalh* and has a greater storage capacity.

The *Ddalh* is carried on the head while *Kilta is* carried on the back and is also provided with two large ropes meant for securing it onto the shoulders much like a bagspack. Raw materials carried in *kiltas* includes fresh apples, pears, potatoes etc. and even farm yard manure (FYM). *Kiltas* are internally lined with gunny bags to provide a cushion so as to prevent injury to the fresh fruit.

For transportation of culled apples, and vegetables like peas, beans, capsicum, potatoes, ginger etc. to distant places gunny bags are used as the packing material.

Packaging of perishables/delicate crop, like tomatoes, is accomplished in bamboo boxes called *Tokra,* which are then covered with thin gunny bags on the top. The shape of the container *Ddalh, Kilta* or *Tokra* provides easy carriage of material on human head or back. The structure is easy to clean and dries soon even after washing it with water. The wide spread use of such containers was probably due to easy availability of bamboo and wood in these areas.

In some parts of Shimla district and Bharmour area of Chamba district, wooden boxes made up of very thick wood were used to pack and transport apples. The weight of the empty wooden boxes ranges from 20 to 25 kg i.e. the empty weight was invariably more than that of apples, it carried in it. The box was made sturdy to enable its carriage on mule backs to distant distribution points/centres.

Packing of Pickles

In lower parts of Himachal pickled mangoes, *galgal,* lime etc.) are packed in earthen pots. The earthen pots are sterilized using fumes generated from burning red chillies along with *Asafoetida* (Heeng) and a little mustard oil. The top of the container is covered with a lid made of wood.

The antimicrobial properties of fumes of red chillies, mustard oil and *Asafoetida* not only sterilize the containers and result in increasing shelf life of its contents but also has potential for preservation which serves as a replacement to inorganic chemicals. An excess use of inorganic chemicals may pose health problems.

Storage of Food Commodities

Food Grains

Food grains like maize, wheat and paddy are stored in special structures made of bamboo called *Peri* or *Peru.*

Prior to use, these structures are plastered on the inside with a mixture of cow dung and clay. These containers are placed on the ground floor and grain is loaded into them from a hole made on the roof of the first floor called *Baurh.* To take out grains, as per need, a special opening is provided near the bottom of each *Peri.* Interestingly, these structures are invariably kept in a separate room called *Overi* and access to which is allowed only to very few persons. The use of bamboo containers allows the free exchange of gases inside the grain and keeping containers on the ground floor ensures cool temperature for storage. Loading from top and unloading from bottom offers easy material handling. Keeping storage structures away from main living room protects grain from fire etc.

In some parts of Sirmour, Solan, Shimla and Kinnaur districts grain is stored in wooden structures known as *Darauntha* and are kept away from *the* main living rooms. *Daraunthas* made of deodar wood are preferred as this wood checks the entry of insects/ larvae.

Wooden houses built away from living quarters check the entry of rodents etc. Windows with wire mesh, provide adequate ventilation. The size of the store house is proportional to the size of land holdings.

Grain storage in earthen rooms: In some parts of Lahaul & Spiti, special earthen rooms are constructed for storing cereal crops immediately after harvest. Storage in such rooms also provides a cool temperature so essential for the storage of grain.

Poor farmers in some parts of temperate Himalayas, have grain storage chambers made of bamboo. These are built out side in the safe vicinity of the farm house.

The logic behind constructing these storage houses/rooms at some distance from the family units was to save the food grains from the hazards such as fire, which was a common feature in the early days, as the entire structure of living rooms was fabricated from wood.

Use of Neem *Leaves/Turmeric/Mustard Oil in Storage*

The use of *Neem* as a pesticide is now well documented. Similarly Walnut, *Bhera* and Mint leaves also seem to possess certain antimicrobial or pesticidal properties, which help in grain storage. The antimicrobial substance in mustard is allyl isothidcyanate. The turmeric powder also appears to perform a similar function. Some of the uses to which these natural products are put are enlisted below:

Grain is mixed with leaves of Neem/Walnut/*Bhera*/Mint (*Pudina*) and then stored in bins.

For storing pulses, specially *Urad,* the grains are mixed with turmeric powder or smeared lightly with mustard oil and then sealed in air tight containers.

Storage of pulses after soaking in cow urine followed by drying is also practised in some parts of the Himachal.

However, the scientific basis for storing pulses after soaking them in cow urine is yet to be explained.

Storage of Tuber Crops

In the tribal belt as well as in some other parts of Himachal, at the on-set of winter, root crops like potato, ginger, turmeric, colocasia (*arbi*) *etc. are* stored in under ground pits and the top is covered with thatch and soil. These products are consumed up to the next summer.

Storage of cabbage heads, meant for seed crop in done, in under ground pits dug in the fields. This is a common practice in Kinnaur.

Apart from other storage benefits, the crop is also protected from the hazards of snow. In Mandi district, a thin layer of grass is spread at the base of the pit to serve as a cushion, prior to storage of the root crops, *the* top of the pit (up to 6") is covered with grass and soil and is raised slightly above ground level to prevent entry of rain water into the pit In some areas, the top of the pit is plastered with cow dung.

Storage of ginger in pits for seed crops is stall prevalent in Sirmour and some part of Solan and Bilaspur districts. The pits in these areas are known as *Khatti.*

These methods of storage provide cool conditions for storage of these commodities, ensuring freshness for prolonged use.

However, it has been observed recently that storage of ginger in pits (*Khatties*) leads to spoilage of almost the entire crop, which might be attributed to a build up of pathogenic inoculum due to long and continuous use of same pits for the same crops. Therefore, digging of new pits is recommended and lighting a fire inside the pit, prior to storage. This is done for creating hygienic conditions inside the pit.

Storage of Fruit. Crops

Citrus Fruits

In Mandi district, ripe *galgal* fruit (*Citrus pseudolimon*) *is* stored in pits dug in the fields. At the base, a layer of ash is spread and then the fruits are placed in alternate layers of ash and fruit. The top of the pit is covered with ash and soil. Ash seems to possess antifungal properties, which help in checking spoilage. Storing in cool pits enhances the shelf life of this citrus fruit.

Another method of storing citrus fruit is the use of a pitcher (earthen pot). The citrus fruit are placed in wide mouthed pitchers and then the pitcher is placed in an underground pit. The top is covered with about a 6" thick layer of soil. Using this technique, citrus fruit can be stored for 25-30 days without any spoilage. This practice is in vogue in the Nurpur (Kangra) area of the State.

Apple

The apples are stored in the tribal areas of Lahaul-Spiti and Kinnaur districts in the underground pits, prior to the onset of winter. Apples are packed into wooden boxes which are then placed in underground pits and the mouth of pits is covered with thatch and soil. In this way, the apples can be stored up to summer time without any appreciable loss in quality.

The method appears to be a modification of the pit storage procedure described earlier, since here the direct contact of fruit and soil in the pit is prevented. A pitcher or wooden box placed in a pit creates a cool store. The development of the zero energy cool chamber (ZECC) appears to have its genesis in this technology.

Banana

Banana fruit wrapped in grass/cloth is also ripened in soil pit and also enhances its storage period.

The shelf life of fruit and vegetables is also enhanced by wrapping them in moist gunny bags through the cooling effect generated by evaporation.

Storage of Gur/Shakkar

In Hamirpur, Una, Bilaspur, parts of Kangra and Paonta Valley of Sirmour districts of the state, it is a common practice to store *gur/ shakkar/jaggery* in large earthen pots commonly called *Rushans.* These *Rushans axe* supported by jute cords to ease under their handling. The top cover is made up of wood, though occasionally, earthen covers are also used.

Another container for packing *Shakkar,* is made up of large sized leaves of the *Torr* tree. Each leaf is held in place by using small bamboo sticks. Each container has a capacity to store about 30-40 kg *gur/shakkar.*

Such containers are called *Purih or Purah.* Such containers, besides being economical, are also completely biodegradable. The exchange of air/gases, permitted in them, helps in the complete drying and best storage of the product.

Storage of Pumpkin

To prevent decay of pumpkin crop and to store it for prolonged periods cow dung is pasted around the stalk of each pumpkin. It is believed that this practice prevents decay and enhances shelf life. However, the true antimicrobial properties of cow dung are yet to be discovered.

Drying of Food Crops

Open sun drying of cereals, pukes, oil seed crops and fruit/ vegetables is an age old practice used for storage of food grains for long periods throughout Himachal. However, the actual methods of drying, show marginal differences in different regions of the state.

Wheat in Chhota Bhangal (Kangra)

Wheat matures in this area during the rainy season when drying of grain proves to be problematic. The wheat and maize cobs are dried using the indirect heat of the hearth (*Chullah*); *fire*. In these areas, residential houses consist of three storeys. The ground floor is meant for sheltering animals while the floor is the residential portion.

Directly above the kitchen, there is a store house where freshly harvested wheat grains are stored. The heat from the *Chullah* below provides the required temperature to dry the wheat grains.

Maize cobs are also dried by keeping them on the rooftop where both kitchen heat and sun facilitate the drying process. The development of the kiln dryer for drying hops in Lahaul Valley and other food crops seems to have its origin from these practices.

Drying of Cereal/Pulses/Oil Seeds

As mentioned earlier, cereals (wheat, paddy), pulses (gram, *wad, masoor,* etc.) and oil seeds (*sarson,* lentil etc.) are harvested when they are just ripe and then kept in the open sun in tile form of small bundles called *Pooli or Poola* for drying. On complete drying, the grain is separated by beating the small bundles against a stone. Due to complete drying, the seeds/grain are easily separated from the husk. However, in present times, threshers and shellers have replaced this practice for wheat and maize respectively. Since completely dry grain is prone to shattering, harvesting crops at the ripe stage, prevents this loss during handling and harvesting.

Drying of Fruit and Vegetables

In the tribal areas and in some parts of Shim la, Sirmour, Kullu and Chamba districts of Himachal Pradesh, fruits such as the apricot, wild apricot (*chuli*), apple and local grapes (in Kinnaur) etc. are dried under the open sun by spreading them out on the roof tops. After 10-15 days of continuous drying, the fruit is collected and packed into gunny bags for subsequent use. Except for the uneven slicing of apples with iron sickles (*Drati* or *Drat*) no pretreatment is administered to any crop. Dried apricot is a commercial product of Kinnaur.

Open sun drying of vegetables like tomatoes, cabbage, turnip (stalk and leaves), mustard leaves is also practised in Kinnaur district and other adjoining areas. After making four incisions on the tomatoes (keeping the slices intact), slicing of turnip (stalks and leaves), cabbage and sarson leaves, these are placed in the open sun for drying. After

15-20 days of continuous drying, the dried products are collected and used during the lean season.

Drying of Unripe Mango Slices

This is a common practice in the lower areas where *in situ* mango varieties are found in abundance. The mature but unripe mangoes are sliced and placed in bamboo baskets (*changer or Chharolu*) which are then kept in the open sun for drying. The dried product called Bukarian is stored in earthen pots and used as an acidulent.

Apricot-drying on the Trees

In the high altitude dry areas e.g. Malling, Nako, Pooh, etc. in Kinnaur district, apricots are not harvested fresh but allowed to dry on the tree itself. Due to very low relative humidity, the apricots dry rather well on the tree. The dried product is of excellent quality, not obtained even after adopting modem techniques of drying (e.g. checking, sulphuring and mechanical dehydration). The apricots dried on the trees are approximately two to three times more expensive than sun dried apricots.

The prevalence of low temperature and dry weather conditions help first in the accumulation of sugar in the fruits and then in the subsequent drying of the fruit to develop a rich colour and sweetness in the product. This is one example of low temperature drying under natural conditions.

Drying of Cucumber and Pumpkin Seeds

The drying of cucumber seeds, using ash is practised in Mandi district of the state. In this method, seeds are mixed with wood ash and then placed on a plate, (*Thali*) in the open sun. The procedure is considered to aid fast drying of seeds since wood ash probably absorbs excess water from the seeds and prevents them from sticking together. The segregation of each seed leads to increase the surface area which results in faster drying.

Drying and Preservation of Meat

In some parts of Shimla, Kinnaur, Lahaul and Spiti and Chamba districts, meat is dried for use during winters, by hanging strips of meat on stings tied across the room. Due to heavy snow, and the low temperature spoilage of the product is prevented. At low temperature and very low humidity, the meat dries very well and is hence preserved better for longer period.

In some places, salt is also sprinkled on the carcass prior to drying, which also assists in preservation.

In Nako and Malling areas of Kinnaur district, the meat is hung in the kitchen. The smoke from the hearth (*Chullah*) serves to smoke the meat, and hence preserves it.

In some parts of Mandi, Kullu and Chamba areas, meat is preserved by smearing it with a mixture of mustard oil and turmeric powder.

Low temperature checks the growth of micro-organisms. Reduction in water activity by drying and use of salt helps in preservation. Smoking is one of the oldest method of food preservation. Both mustard oil and turmeric powder are known to possess antimicrobial properties, and thus help in preserving meat.

Dried Ginger (Sonth)

In Shillai and Renuka areas of Sirmour district, ginger is also sun dried. Fresh ginger upon soaking in water is rubbed against a gunny bag or placed in bamboo baskets to remove the peel. The coarsely peeled ginger is then dipped in lime solution followed by warmth for faster and better fermentation.

Distilled Country Liquor

Preparation of alcoholic beverages by distillation is common throughout the tribal belt of Himachal Pradesh. Government excise department too permits this practice, allowing each family to keep up to 24 bottles in this possession at any given time thereby acknowledging the socio-culture fabric of the region.

Fermentation

The liquid is prepared from fruits like local grapes (coloured and seeded), apricot, wild apricot (*chuli*), pear, apple, wild almond (*bemi*), etc. In higher reaches of Kinnaur, however, millet is the main ingredient. Country liquor is a strong alcoholic drink obtained after distillation of fermented musk.

Fruits like apple or pear are cut into coarse pieces, grapes, *chuli* and *apricots* are either hand washed or washed by feet and then placed into large tanks (concrete or wooden). In the agricultural lean period, dried fruits like apple slices, grapes and *chuli* are also used for this purpose. Generally wooden drums are preferred for fermentation as they maintain suitable temperatures. To obtain higher alcohol output, occasionally molasses are also added to the fermenting musk. Drums are then tightly sealed with wooden or heavy stone lids (Weather permitting, the fermentation continues for 15-20 days). The completion of fermentation is judged using different parameters like alcoholic or acetuous odour, small vapours appearing on the inside of the lid or non-sticky nature of the pulp, when pressed between the forefinger and the thumb. All these characteristics are indicative of complete fermentation.

In some places the fermenting vat or bottle is placed in a cow dung pit (*Mahlash*). Generation of heat during decomposition of cow dung in the pit provides required warmth for faster and better fermentation.

Distillation

The fermented musk is then distilled to obtain the liquor by different methods.

The fermented liquid is boiled in a vessel called as *Baltoi*. At the base of which a temporary stand is made under place a Container to collect distilled liquor. Upon heating, the fermented liquid boils, and its vapours are cooled by striking the top of the container (holding cold water), the condensed vapours drop into the container, meant for collecting the liquor. Cold water is repeatedly replaced until the process is completed.

In this manner, recovery is comparatively low since some vapours also fall back into the boiling liquid. Moreover, there is no provision for complete separation of its methyl alcohol content.

In some places, this method has been further improved as under: Herein, an empty *ghee* tin (canister 15 kg capacity) is made slightly rounded, and at its base, is placed a temporary stand on which a pot is placed for collecting the condensed vapours. The pot is attached to a rubber pipe to receive the liquor in bottles/plastic cans. This has a higher efficiency and it also allows the separation of the initial drops of the liquor which contain high levels of methanol.

As a further improvement in the distillation process, the fermented musk is placed in a metallic pitcher which is covered with a (stone) slate with a hole at its centre for the passing out vapours. The slate is used ostensibly to avoid the overflow of the material as it boils. Another metallic pan with a hole is kept on the slate and is attached to a side pipe where condensed alcoholic vapours are collected. To cool the vapours, this container is housed in a bigger metal pan which is cooled by continuously running tap water. Two rubber pipes viz. inlet and outlet aid the circulation of cold water and the release of warm water. The pan is maintained in a tilted position, so that the inlet of the pipe is located higher than the outlet pipe. The bottom of the utensil is cooled, thus the alcoholic vapours condense on its lower surface. The condensed material runs out through the side pipe of the lower utensil from where it is collected into bottles or plastic cans. Approximately, 12-14 bottles of alcohols are collected from 40 kg of fermented musk. However, by using 5-6 kg of molasses/*gur* in this musk, the yield increases to 18-19 bottles. Fuel wood is used for heating. All joints are secured using barley, *phafra* or wheat paste. In some places even dung paste is used for securing joints.

The simplest method involves a rubber pipe being inserted through a hole in the slate lid, to collect the vapours. The vapours are cooled by maintaining a continuous flow of cold water from the tap. The condensed alcoholic vapours are received in the bottle.

Improved Method of Distilling Country Liquor

In some areas of Himachal the above stated methods have been further improved by replacing the water trough by rubber pipe, to collect vapours from the boiling fermented liquor. In this method an empty *ghee* tin i.e. 15 kg capacity canister, is used for boiling the fermented liquid by heating it on fire. On the top opening of the canister, a rubber pipe is inserted to receive the alcoholic vapours, the pipe is paired through a water trough to condense the vapours. The alcoholic product is then collected in glass bottles or plastic cans.

In order to increase the alcohol yield Gur is sometimes added. In the agricultural lean period, dried grapes, apple and *chuli* etc. are used for preparing the fermented musk. The product prepared from grapes is called *Angoori*. Chhang is prepared using Millets and Ghanti is prepared from any other source. During initial distillation, first 1-2 bottles of liquor are called Moon and contain the maximum amount of alcohol while last distillate contains comparatively lower amounts of alcohol and is called *Rashi*. Costwise, Angoori is the costliest, followed by Moori and least expensive is *Rashi* often only Rs. 10-20 per bottle (700 ml).

Due to extreme winters, the consumption of country liquor (*ghanti*) is extensive in the entire cold dessert regions for medicinal purposes. Besides social use the product is also used during rituals and religious occasions.

Traditional Apple Plucking

It is done by climbing an apple tree with putting by a small basket made, which is put around the neck of gunny bag cloth. This prevents the damage of fruits. Plucked apples, from the gunny bags are transferred to a bigger basket (Kilta) which is also provided with a rough cloth cushion. In this way, apples are carried to the store room for grading.

Steam Cooking

A big special vessel made of brass is used for preparing steam cooked cake of maize or barley. This vessel is called Chalta in which at the bottom of the vessel water and rice straw is placed. Above this round cover, made from maize or barley flour is placed and cooked in steam.

Fruits of Arid and Semi-arid Areas

Date-palm. (*Phoenix dactylifera* L.).Some of the imported varieties of date-palm have been cultivated successfully in Punjab. This tree thrives in light as well as in heavy soils, provided the depth is adequate (2.5 to 3 metres) and the drainage is good. The plant is not affected by frost and requires intense heat in summer for the development and ripening of the fruits. The maxim 'head in fire and feet in water' indicates the conditions that are ideal for date-cultivation. A wet season during flowering and fruit-ripening is limiting factor in the successful cultivation of the date-palm. South-western Punjab, northern Rajasthan and Kutch offer suitable conditions for date growing.

Climate

The date palm is a tree flourishes well under varied sets of climate. However, for proper maturity of fruit, the date requires prolonged summer heat without rain or high humidity during the ripening period. A light shower with prolonged periods of cloudy weather and high humidity may cause more damage than a heavy rain followed by clear weather and drying winds. The mean temperature between the period of flowering and ripening of the fruit should be above 21° C rising to 27° C or higher for at least one month. For successful fruit maturation, nearly 3000 heat units are required. The heat units available in most of the north-western districts of India, counted from the time of flowering, i. e. end February to July, indicate that these are suitable for date palm cultivation. Taking this into consideration, parts of Rajasthan extending from Sri Ganganagar to Anupgarh, Jaisalmer and Barmer and Kutch district of Gujarat will be suitable for date cultivation. The climate of Abohar (Panjab) should be considered marginal because of rainfall (30 cm) and high humidity at the time of ripening (July-August) being the limiting factor.

Soil

The date palm can be grown in soils containing more alkali or salts. It can tolerate such condition better than many other fruit plants. In view of the large investment required to bring a date garden into bearing and maintain it on profitable production, sandy loam soil, 2 or 3 metres deep with good water holding capacity and drainage is most desirable. Date palm can tolerate high soil salinity. In the soil having 4 per cent salt concentrataion, dates can survive well provided the root system does not come in contact with a stratum of soil where the sodicity is more than 1 per cent.

Choice of Varieties

Hillawi : It is soft date from Iraq. This variety is relished both in doka and dang stages. Total soluble solids range between 28 to 42% and astringency in the fruit at doka stage is low or almost absent. It is an early variety yielding good crops and is somewhat more tolerant to rains. Dry dates (Chhuhara) of good quality and cured soft dates (Khajoor) can be prepared. Its fruit at dang stage (fresh form) is very delicious. Average yield ranges from 50 to 80 kg per palm.

Khadrawi : It also originated from Iraq. This variety proved successful under Abohar conditions. The trees of this variety are comparatively less tall and and yield good crops. It can be used both for dry dates and soft dates. The yield ranges from 40 to 70 kg per palm.

Shamran : It is a mid-season cultiver tolerant to high humidity. The variety yielding prolific crops. The fruits can be cured into good quality. Chhuhara and also softened with salt treatment. Fruit is medium to large, oblong to oblong-oval and yellow at doka stage. Its yield per palm is 40 to 70 kg.

Medjool : The variety originated in Morocco and has large size fruits. It is late in ripening and has proved particularlt good in preparing dry dates of attractive bold size and good quality. Its yiled is 35 to 50 kg per palm.

Barhee : It is soft date from Iraq. The doka fruit contains about 32% TSS, has golden yellow colour and has a very pleasant taste because of low astringency and high pulp content. The trees of this variety are prolific in yields and the fruit is ready for eating late in the season i.e. first fortnight of August at Abohar. Its yield ranges from 60 to 110 kg per palm.

Hayany : It is soft date originated from the UAR. The berries have attractive deep colour. The fruits could be consumed as fresh in the doka stage as ripening does not proceed further under Abohar conditions. The yield per palm ranges from 30 - 40 kg.

Zaidi : It is a mid season variety, slightly tolerant to rain or high humidity. The fruit is small to medium, obovate and yellow at doka stage.

Propagation Techniques

The date palm are always propagated throughsuckers (offshoots) for commercial plantation. The suckers usually arise near the ground around the trunk. The offshoots arising at a higher level from the ground on tree trunk and with no roots give very poor survival. These are separated from the mother palms during March or August-September. The leaves are trimmed off 4 - 5 days before their detachment. The inner leaves should be cut back by one-half and the outer ones by two-thirds. The tender young unopened leaves near the central bud and parts of the bare stalks of the old leaves necessary to protect the bud, should be kept on the offshoots. The outer whirl of leaves may be fastened to afford protection from heat and cold to the central bud also facilitate detachment and transport.

The suckers weighing approximately 25 to 30 kg and well rooted should be removed. Expose the point of detachment and set a chisel (12 cm wide and 19 cm long) sharpened at the face and welded to a 150 cm long handle of 4 cm diameter. Drive in the chisel by hammering at the other end. It is very important that no injury is caused to the mother palm. Cut surface should by coated with tarcol or some suitable material immediately and earthed up.

Planting Operation

The offshoots are planted 6 to 7 metres apart in the already prepared pits of 1 * 1 * 1 m size immediately after detachment. In this way 202 to 275 plants/hectare will be accomodated. If transported from long distance, the offshoots may be kept in the straight. The soil around the offshoots should be pressed firmly. The field is irrigated immediately after planting. Thereafter, frequent light irrigations are given to keep the soil always moist. At Abohar, the offshoots removed in March survived better in direct field planting than those removed in August-September. However, the separation could be done both in February-March and August-September.

Orchard Cultural Practices

Irrigation: Date palm is known as drought resistant fruit tree and is able to survive for long periods without irrigation. However, continuous drought condition retards the growth of the plant. If water is available, date palm uses it lavishly. Date palm is highly tolerant to excessive irrigation and floods. Continuous stagnation of water or waterlogged conditions are injurious for its growth. To maintain maximum growth, the root zone up to 2 - 3 m should be kept moist and not allowed to dry. Light but frequent irrigation should be given after planting.

The full grown trees on the lighter soils are usually irrigated 7 to 10 days during mid-summer and every 15 to 30 days during winter. Such soils requires 2.75 to 3.75 acre-meters of water per year and 30 to 45 acre-centimeter per month is needed during summer. On heavy soils, half the amount may be enough. Each irrigation of 10-15 acre-centimeters is usually sufficient. Irrigation is withheld when ripening starts to facilitate harvesting, hasten fruit ripening and to reduce fruit drop caused by high humidity.

Manuring and Fertilization

Generally the farmers do not apply manure to the date palm. The trees certainly respond to manuring as indicated by increased vigour and growth. The results of studies conducted at Abohar shows that 25 to 50 kg of farmyard manure and 1 to 2 kg ammonium sulphate per palm should be added. The quantities may be reduced if some leguminous crop has been grown in the date palm. Farmyard manure is applied during December-January. The nitrogen dose should be given a fortnight before flowering i.e.in the first week of February.

Intercropping

Intercropping in date palm with suitable crops bring good income and also improves the fertility of the soil. During the first few years, intercropping can be practised with no shortage of irrigation. Intercrops such as gram, peas, mash, moong, moth, senji and lentil can be sown during summer. Intercropping of some vegetables in plantation located near the cities, can be practised if sufficient irrigation and manuring facilities are available. The filler trees like grapes can be tried with good success.

Fruit Thinning

Excess load of fruit may cause shrivelling of berries, breaking of spathe stalks, more damage due to rain and humidity, delay in ripening and alternate bearing. It also reduced size and produce poor quality of fruit. It is, therefore necessary to keep only optimum quantity of fruit and thin out the rest. This is usually accomplished either by reducing the number of fruits on each bunch and or by removing some of the bunches. The number of fruit that a palm can safely carry depends on the cultivar, age, size and vigour of the palm and number of green leaves on it. Under normal conditions, 1 - 2 bunches in the 4th year and 3 - 4 bunches in the 5th year may be left. Normally 8 - 10 bunches per palm are retained in India. Small, defective and broken bunches should be removed.

In short-stranded varieties like Khadrawi, the strands are generally cut back to even up the bunch from the top. Most of the fruit thinning is done by the removal of half to two-third of strands from the centre. In the long stranded varieties like Deglet Noor, one-third to half strands are cut in similar way as in Khadrawi, in addition, strands are also cut back to remove about one-third of the flowers.

The desirable number of fruits to be left is between 1300 and 1600 per palm depending on the variety. The per cent thinning is generally done 40 - 50 in Khadrawi, 50 - 55 in Hallawi, 50 - 60 in Zaidi and Barhee. Ethephon 100 - 400 ppm after 10 to 30 days from fruit set was found effective in fruit thinning of cv. Hayany. The biennial bearing habit of the treated palms were found to reduce by ethephon treatment. It also advances the ripening of fruit.

Arid Zone Agroforestry: Dimensions and Directions for Sustainable Livelihoods

Drylands cover more than 60% of the earth's surface. Arid zones are described as a part of the drylands, and have the most severe edapho-climatic conditions. The major distinguishing feature for defining and for planning for the arid zone is the low rainfall (below 500mm or Aridity Index <0.20) with more than 50% inter-annual variability. This makes a great difference in terms of the nature of the ecosystem, the socio-economic environment and the challenges for sustainability. High wind and solar regimes further increase the effect of rainfall variability and the whole complex makes a fragile ecosystem in which small disturbances may cause great loss to the sustainability, which are sometimes irreversible (Houerou, 1996). The hot arid zones of the world are economically and environmentally disadvantaged, with unique problems. The ecosystems of these zones are highly fragile with large risks that cause severe impediments to development programs. Of the total land area in the world, arid zones cover 18.8%. The hyper-arid zones account for 22.3% of the total arid zones (FAO, 1989).

These arid zones are diverse in terms of climate, soils, vegetation, animals, and the lifestyles and activities of the people. Little but variability in rainfall and the presence of distinctive periods of drought are the characteristics of the arid tropics. Often, the terms drought and aridity are used incorrectly. A drought is a departure from average or normal conditions in which shortage of water adversely impacts on the functioning of ecosystems, and the resident population of people, whereas aridity refers to the average conditions of limited rainfall and water supplies, not to the departures there from. In general, arid zones are characterized by pastoralism and little farming but there are always exceptions (Malhotra, 1984).

Arid zones are found in almost all the continents of the world (Table 1). In North America, arid zones are found in the southern US and Mexico, and in South America they are located in Brazil, Argentina, Chile, and in some other countries. In Africa, substantial parts of Saharan Africa, Ethiopia, and Namibia, are classed as arid environments.

The hot arid regions of India lie between 24° and 29° N latitude, and 70° and 76° E longitude, covering an area of 31.70 million hectares, and involving seven states: Rajasthan, Gujarat, Punjab, Haryana, Andhra Pradesh, Karnataka, and Maharashtra. In total, 11.8% of the country is under a hot arid environment. The arid regions of Rajasthan, Gujarat, Punjab, and Haryana together constitute the Great Indian Desert, better known as the Thar Desert, which accounts for 89.6% of the total hot arid regions of India, and constitutes the principal hot arid zone of the country.

In arid zones, vegetation is typically sparse, and is comprised of perennial and annual grasses, other herbaceous plants, shrubs and small trees. The native plant species have adaptations that enable them to reproduce, grow and survive in the most inhospitable edapho-climatic conditions. Some plants have evolved special root systems, while others have unique leaf characteristics that allow them to withstand prolonged periods of drought.

The number of tree species is very limited in arid zones, and in general, they are very slow growing due to limitations of environmental conditions, but nowhere in the world are they are so intricately associated with the life of human beings.

To evade or minimize the adverse affects of frequent droughts, the native peoples in arid zones have often developed production systems in which woody perennials have a very important role, both from a productivity as well as a resource conservation point of view (Tewari, 2000).

Their centuries-old experiences, under diverse socioeconomic and cultural conditions, were passed from generation to generation and people have established well-contained systems of production (silivipastoral/agrisilvicultural/agrisilvipastoral/livestock husbandry) that accrue maximum benefits from their woody components. When arid zones have been looked at from the perspective of forestry development, the focus has been on the management of trees and shrubs that are either native to a particular arid zone, or have been introduced, especially for conservation purposes.

However, concern for trees and forests has increased in arid zones as it has in many other regions in recent years. This concern has been fuelled mostly by the disappearance of large areas of woodlands and trees from such fragile ecosystems in many countries of world. In the arid zone, closed forests are seldom available. In India, traditional animal husbandry and agroforestry practices have been used to manage parklands, rangelands, reserved silvipastures near holy places, lay farming, and run-off farming (traditional watershed management).

Trees are managed mainly for their non-wood forest products (NWFP) and their environmental services. Animals are an essential part of the production system. The whole system has been traditionally developed to spread the risk of drought in diversified components and efficient utilization of scarcely available natural resources.

Arid Zone Agroforestry

As stated above, the prevalent form of forest management in arid lands is agroforestry, which is concerned with the management of trees and shrubs for conservation and sustainable development. Agroforestry in arid lands differs in many ways from the timber forestry practiced in moist tropical and other temperate ecosystems and in many cases its applications are broader in their scope. These applications include production of wood for fuel and poles; horticultural practices for fruit and nut production; range management for fodder and forage production; forestry to modify microclimate for improved agricultural crop production; and protection forestry on land susceptible to wind and water erosion. Many rural people combine these applications of agroforestry with other land uses linked to their needs and social values. The value of forestry in the arid zone is difficult to evaluate in monetary terms. Arid zone agroforestry therefore needs to be understood broadly as management of trees and shrubs to improve the livelihood and quality of life for rural people in the arid environment. It is in this diverse context that policy developemnt must take place if sustainable development of forest resources in the arid zone is to be achieved.

Agroforestry is being followed traditionally in the arid zone to minimize the risk of drought and sustain life in climatic adversities. Several need-based systems have developed traditionally and some have been improved or incorporated in recent times.

Trees in Life Support Systems— Traditional Agroforestry

The image of the desert as little but vegetation-less, unbroken, stretches of sand and inaccessible terrain conditions is shattered upon entering the arid zones. There are sparsely distributed trees, beneath which are cultivated arable crops (especially during the *kharif* season, as agriculture is predominantly rain fed) and/or grasses, and other herbaceous flora, in long stretches interspersed among widely scattered village settlements and *dhanis* (a unique settlement pattern of the Thar intended for the life of agriculturists' families during the active cropping period away from the village, but nearby their fields). The Indian arid zone is a vegetated desert having a very rich floral diversity of about 682 species, out of which 131 are known for their economic uses (Bhandari, 1990).

Farmers in the hot Indian arid zone grow arable crops in association with tree species. In fact, such integration of arable crops with trees in the farming systems is a unique, combined, protective-productive system that works on the principles of ecology, productivity, economics, and sustainability. These systems are now generally referred to as agroforestry. Because most trees are drought resistant, they are still able to provide fuel, fodder, fruit and other products, when the crops fail, as frequent droughts are a common phenomenon. Thus, trees have a very important place in the life of people in the arid zone of India, as they are directly related to the livelihood of inhabitants, and also provide the important service of climate moderation in an inhospitable environment (Sharma and Gupta, 1996).

Run-off Farming

This is a system of growing crops on harvested and stored water in the farm soils by constructing an earthen dam or a bund across the gentle slope of the farmland in the valley bottom. Shallow, gravelly and rocky uplands normally used for grazing are used as catchments for efficiently harvesting the runoff water. This is locally known as the Khadin system. The major components of the Khadin system are the water collecting area (low rocky catchment), contour bands (channels), moisture storage basins (khadin farm), water impounding mechanisms (bands, spillways and sluice) and a zone of cultivator's settlements.

In this region monsoon rainfall is scanty, erratic and generally the water requirement of most of the crop plants is not fully and properly met. With the result, cropping in the Kharif (rainy season) often becomes a gamble. To offset such a risk and to have some assured crop yield in such environment, one very thoughtfully developed traditional system known as Khadin, has been followed in this region for centuries (Kolavalli and Whitaker, 1996). This system makes the use of shallow gravelly and rocky uplands as the catchments for efficient harvesting of runoff water, which otherwise normally used for grazing. The water is trapped in the low lying valley plains that are converted into farmland where either the kharif or the rabi (winter) crops are raised, depending upon the amount of rainfall and consequent runoff received during the monsoon. The system takes into account not only the vagaries of the climate to raise the crop with such a low rainfall but also prevent any soil deterioration.

Silvi-pasture

In the arid zone there are areas where rainfall is below 200 mm/yr and with this rainfall food production is very difficult. On the other hand there are some grass species, *e.g. Cenchrus cilieria, Lasiurus sindicus* etc., which are very well adapted to such climate and make natural rangelands. The climax tree species like *Prosopis cineraria* and *Zizyphus numularia* naturally come up in these rangelands and make a silvipastural system. Animals like cows, goats and sheep are the part of this farming system. Traditionally the areas near holy places like temples are kept reserved for silvipasture with very well established social fencing. This is a very good example of sustainable management of common property resources that still exist in several places. The silvipasture systems are strictly used according to the land's carrying capacity. Also these systems cover large areas i.e. 100 to 500 km^2. These systems have low productivity but are an example of an excellent synergism between the nature and the natives and that's why they have been sustained for centuries (Shankarnarayan *et.al.* 1987).

Agri-silviculture

This is the most popular system in the areas receiving rainfall between 200-400 mm/yr. People protect the naturally germinated seedlings of useful trees like *P. cineraria*, *Z. numularia*, *Tecomela undulata*, etc., which come up in the crop fields. Mixed cropping of pearl millet, moth bean, cluster bean and sesame is carried out under these trees. These trees do not compete with the crops but rather are complimentary in terms of improving the microenvironment. These trees give fodder, fuel, fruit etc. even when crop fails during drought, thus provide sustenance to the farmer during drought. The *P.cineraria* and Pearl millet agroforestry is well known around the world. Some other aspects of this system are discussed with the Non wood forest products.

New Systems Developed

Some of the agroforestry systems have been developed after research and on farm testing are discussed below.

Agri-Horticulture. After two decades of experiments on research farms and farmer's fields, a new agri-horticulture system of jujube (*Ziziphus mauritiana*) inter-cropped with arid legumes *e.g.* clusterbean/ mothbean/greengram has been developed for the areas receiving rainfall of more then 250 mm. This agri-horticulture system has been found to give better and earlier production, year-round work and resilience to erratic rainfall. In this system initially the jujube saplings are raised in the farmer's nursery by stored rainwater (farm pond) and then saplings are transplanted at a distance of 5 x 10 meters irrigated twice in first two weeks; thereafter the trees receive only harvested rainwater. Arid legume crops are raised successfully in the interspaces of jujube. It is a short duration (50-75 days) crop, sown at the onset of monsoon and harvested in September/October, *i.e.* before the full growth and blooming of jujube. This offers least competition and both are complimentary as the jujube provides a favorable environment for arid legumes while legumes fix nitrogen, and leaf fall adds organic matter to the soil. Besides fruits and wood from pruning (for fuel or fencing), leaves as fodder are the additional products from jujube. Overall productivity of the system in normal and drought years shows less variation in yield due to variations in rainfall. This system can provide economic returns of Rs 28000-37000/yr as compared to the sole cropping of arid legumes (Rs. 8000-12000) with low input supply, besides non-tangible environmental improvements which give resilience to drought.

Lay farming. In this system, on one piece of land a rotation of grasses for 4-6 years, followed by food grain crops like pearl millet or legumes for 2-3 years is followed and then the land is left fallow for 2-3 years. The whole farm is divided into parts in such a way that every year all three practices, *i.e.* grass production, crop production and fallow, are available in one or another part. In this traditional system the rotation was designed considering the poor carrying capacity of arid soils. For that purpose first grasses are grown which improve the soil's physical and biological conditions, then good production is obtained, followed by a fallow period for natural restoration of the soil. During the fallow period several weeds come which improve the soil and the life cycle of various pests is broken. This system is gradually vanishing because of several social changes which are based on the thinking of exploiting nature rather working in synergism with nature.

Wind breaks/shelterbelts. Wind erosion, a high thermal regime and hot desiccating winds are some of the serious problems, which affect the establishment, growth and yield of crops in arid areas. A mixture of trees and shrubs planted across the wind direction help in reducing the wind speed. Evaporative demand studies conducted at CAZRI with three row shelterbelts of *Acacia tortilis*, *Cassia siamea* and *Prosopis juliflora* as side rows and *Albizzia lebbeck* as the central row were found effective in reducing wind speed and wind erosion. Micro-shelterbelts of tall growing plants have also been found quite effective in providing protection with about a 20 to 40 per cent increase in yield of vegetable crops grown with irrigation.

Challenges for Sustainable Arid Zone Agroforestry

Agroforestry has been followed since time immemorial for sustainable livelihood in the arid zone, but in the recent past some changes have occurred, due to which the need is being felt to develop agroforestry systems, which may be able to meet the requirements of sustainable livelihood in the future. Some of these changes are discussed here.

Heavy Biotic Pressure

Chalcolithic and Iron Age maps of the Indian subcontinent suggest that human occupation of the current Thar Desert region was thin during these earlier ages, and that populations depended upon hunting and limited pastoralism. From the analysis of available census records, the population of the Thar Desert region registered an increase of 490% from 1901 to 1991. The present population of the Thar is about 40 million. Decennial variation in population from 1901 to 1991 in the Thar showed a growth rate of 186% (between 1901 and 1971) as compared with 132% for the whole country. The population growth rate for the decades 1971–1981 and 1981–1991 were 36.7% and 30.7%, respectively. The population density is quite high compared with the global average of 6–8 persons/km^2 for arid zones. The major factor responsible for such phenomenal growth of population in the Thar is a wide gap between birth and death rates. Moreover, social values favor large families and positive sanctions for fertility outnumber negative ones.

Cropping on Marginal Lands

Of the total area of the Thar, only 1.3% is now under forest cover. The net cultivated area increased by 54.1% from 1915 to the early 1980s. This trend in cultivated area increase has further accelerated during the past decade (the 1990s), having increased by 11.7% compared with the amount in the early 1980s. As a consequence of this, barren, cultivable and uncultivable wastelands, permanent pastures, and fallow land have declined substantially (Dhir, 1997). There was a significant decline in grazing land area as marginal lands were brought under cultivation in recent years (the late 1990s). Because of low and erratic precipitation, restricted to a very short period, only one crop, the kharif (monsoon) season crop, is sown each year in most of the Thar. The common crops are pearl millet, cluster bean, green gram, moth bean and sesame. In the arid zone of peninsular India, in addition to these crops, sorghum, maize pigeon pea and other coarse millets are also grown. It has been observed that the agricultural holdings during late medieval times were only one fifth of those found today. By 1951, much of the potential agricultural land was brought under plow, and the trend since then in already established agricultural tracts has been intensification of farming, and the expansion of agriculture into new lands that may be considered environmentally marginal. Mechanized field operations (use of tractors) not only increase soil erosion but also reduces the naturally regenerated population of various useful species (Jodha, 1997).

Unsustainable Development Policies

With the advancement of science and technology, efforts have also been made to enhance production in the arid zone. Some of the technologies like sand dune stabilization, wind-breaks and agri-horticulture have produced good results because they are developed in synergism with nature (ecosystem) and natives (people and vegetation). But some other technologies imported from less environmentally stressed areas have proven to be unsustainable. For example use of tractors adversely affected the natural regeneration of native trees like *Prosopis cineraria*, and increases wind erosion. Irrigation with waters from the Indira Gandhi canal network has caused problems of water logging and, secondary salinization (Dhir, 1983). Several exotic species either become invasive, like *Prosopis juliflora*, or didn't prove economically productive, like *Faidherbia albida*, jojaba etc. The overall impact of these imported technologies, along with high biotic pressure and a fragmented approach to desertification control, can now be observed in terms of loss of biodiversity, increasing desertification, and other adverse impacts. It is now imperative to think broadly and plan holistically for the future of arid zone agroforestry, the basis of sustainable livelihood in the arid zone. Throughout the arid zones, there is no dearth of problems, but rapidly increasing desertification (some call it land degradation) is a worldwide problem. Land degradation appears to be more of a descriptive term than desertification, as it better describes the process of wind and water erosion, soil salinization, and loss of vegetation cover. It helps in clarifying relationships among climate, topography, soil structure, land use, hydrology, and vegetation. Throughout the arid zones of the world, the major factor responsible for land degradation is soil degradation. At the world scale, the GLASSOD database indicates that in arid zones, more than 60% of soil degradation is caused by wind erosion, with human-induced water erosion accounting for 8% of the phenomenon. At the global level, 349.6 million hectares of land in arid zones are characterized by a light to moderate degree of soil degradation, and 42.9 million hectares have strong to extreme soil degradation. Besides wind and water erosion, over-grazing, high magnitude agricultural degradation activities, over exploitation of vegetation for domestic use, *e.g.* for fuel, and faulty land use are other factors responsible for ever increasing land degradation in arid zones.

The links between land or soil degradation and vegetation in arid zones are complex. Vegetation and soils display differing resilience to disturbance. Although in arid zones vegetation communities can be readily disturbed and degraded, recovery rates are very slow in comparison with those of other ecoclimatic conditions. Vegetation in arid zones is highty sensitive to natural climatic variability. Decreased production and an ability to lie dormant are two strategies displayed by arid zone plants for survival at times of moisture deficiency. Although an adequate database is not available, two useful relationships between soil degradation and vegetation can be determined. First, vegetation cover may help to indicate the vulnerability of a particular habitat to soil degradation, and second, vegetation cover or community type or plant species may indicate that soil degradation.

Policy, Research and Development Recommendations

Every ecosystem has some buffering capacity that enables it to remain unchanged for some time if it is overexploited. While this applies to the arid zone, its fragile nature limits its buffering capacity. Thus, for a few decades the shortcomings of policy were not observed but as the buffering capacity is gradually being eroded, the indicators of unsustainability as a whole, *e.g.* decreasing natural resources, deforestation and desertification, become visible. For fragile ecosystems, like those that occur in arid zones, one thing needs to be recognized by policy makers and scientists, that is; sustainability of agroforestry in the arid zone depends on the degree of synergism with '3Ns' i.e. Nature, Native vegetation and Native people.

Some of the future directions for arid zone agroforestry need to focus on the evaluation, improvement and encouragement of sustainable traditional agroforestry systems, exploring the possibilities of natural regeneration/rehabilitation of arid zone ecosystems, NWFPs as the basis of sustainable livelihood, rethinking of research priorities, and policy improvement. These will be discussed in turn.

Evaluation, improvement and encouragement of sustainable traditional agroforestry systems that have been developed for centuries in harmony with nature. As mentioned earlier, agroforestry systems have been developed and used in this region since time immemorial. This may be because the inhabitants of this region found that no other option was sustainable aside from conserving and harvesting whatever nature is giving. The present forms of traditional agroforestry developed over the centuries through the experiments of earlier generations in all types of climatic variations. The validation of their sustainability is very difficult to establish by conducting short-term experiments lasting only 5-6 years. Thus, there is immense possibility for evaluation, improvement and encouragement of these traditional agroforestry systems.

Exploiting the possibilities of natural regeneration/rehabilitation of arid zone/arid zone agroforestry as an alternative to failed plantation/rehabilitation strategies. It is a mechanism of restoration or maintenance of ecosystems by nature that trees produce numerous seeds. A small part of this seed population is sufficient to replace the loss of plant population due to various factors. Human beings propagate the plants artificially to compensate for what they harvested form the natural vegetation or to get a higher quantity of the natural products. Those needed in large quantities have already been domesticated, *e.g.* food crops, fruit trees *etc.* Still a large number of natural species are wild or depend on natural regeneration. Trees in agroforestry may fall under this category. Natural regeneration as well as artificial tree establishing technologies works comparatively better in humid and irrigated areas. In the arid zone, due to climatological, physiological and sociological reasons artificial establishment has not achieved the required success.

The need for domestication and technology development for sustainable harvesting and value addition for non-wood forest products (NWFPs). Unlike the humid or sub-humid regions where trees are typically grown for timber, the trees are grown in this arid region mainly for non-wood forest products (NWFPs). In the arid zone closed forests are seldom available and agroforestry has been followed since time immemorial in which people conserve and manage whatever seeds of native tree species like *Prosopis cineraria* and *Zizyphus numularia* germinate in the crop field or on the common land. After the establishment of trees only annual lopping of branches is done which provides precious fodder during scarcity/drought. Taking the example of *P. cineraria*, this tree is found in this region at densities ranging from 40 to 120 trees/ha. This we can say is the natural forest in common or private lands otherwise no dense forest is found as in the high rainfall areas. People lop the branches of this tree in the month of December and it becomes lush green in the month of May-June when no other fodder is available. Simultaneously, pods develop which are used as vegetables. Similarly other tree species provide fodder, pods, fruits etc. The interesting thing is that, during drought when annual crops fail, the yield of NWFPs does not decrease drastically, thus trees provide food security during drought. Besides some trees produce other products like gum, resins, edible/non-edible oils *etc.* which increase the income of the marginal farmers who are in the majority in this region (Sharma, 2000).

The services provided by the trees in hot arid regions are equally important for moderating the climate. Cultivated in the form of shelterbelts, they reduce wind erosion and conserve soil and water. In agroforestry systems, trees reduce the solar heat load on crops and also decrease wind speed and soil temperature, conserve soil moisture, increase biological activities in the soil and thereby improve the microenvironment for crops. The yield of crops grown under *P. cineraria* is generally higher than sole crops, knowing this fact agroforestry has been followed in this region for centuries. The edapho-climatic conditions in this region are so harsh that every tree improves the environment, it may be at micro level, yet its impact is visible (Sharma, 1998).

The other important NWFPs in these arid areas are plants with medicinal values. Due to climatic specialty there are medicinal plants which are either endemic to this region or their quality deteriorate if they are grow/found in higher rainfall areas.

Research priorities. Considering the social and ecological demands, policy on agroforestry research also needs to be prioritized. Some of the possible priorities are:

In-situ rainwater conservation for tree establishment and for agroforestry more generally. Water is the most limiting factor for productivity and sustainability in the arid zone. As described earlier, the effect of bringing additional water through canals or ground water has been the cause of unsustainability in the ecosystem. Thus the only possibility remains the conservation of rainwater (Gupta and Sharma, 1998). Trees are the keystone of arid zone conservation and once trees are established the other conservation processes start automatically e.g. rainwater conservation, development of vegetation cover etc. (Sharma, 2000).

Further research is required on:

(i) Development of *in-situ* rainwater conservation techniques that require less labor and maintenance;

(ii) Development of plant-based water absorbing/retaining materials;

(iii) Initiatives for widespread adoption of rainwater conservation techniques;

(iv) Improvement in traditional water harvesting systems;

(v) Development of simplified windbreak establishment techniques.

Improvement of policy. There are some important aspects on which improvement of policy is needed for creating a more congenial environment for development. These include:

Increased international funding for arid-zone research and development. The arid regions of the world in general suffer from a vicious cycle of low productivity, a low level of investment and as result, poverty. Greater priority is given at the international level for funding of projects and programs related to timber forestry and rainforest management, perhaps because these activities contribute more to international trade. Very limited funds aimed at combating desertification are available to the arid lands. Forestry and agroforestry activities in arid regions no doubt contribute less to international trade but they provide substantially support to the livelihood of people in these climatically less favored regions. There is a need to rationalize these priorities and for the international community to give due importance to the non-tradable benefits of forestry/agroforestry, which are otherwise the basis of sustainability and livelihood.

Suggested areas for increased investment in the arid zone are as follows:

(i) ***Organic farming in virgin arid lands:*** The philosophy of the traditional systems was to work in synergism with nature and use locally available natural resources. High input agriculture (HIA) on the other hand is only sustainable with assured input supply, *e.g.* irrigation, fertilizers *etc*, which is not always possible in drylands. Thus the chances of failure are much greater than in traditional systems during drought. Also due to less use of synthetic chemicals in the past, these lands are comparatively free from the residues of such chemicals. Hence organic farming, based on traditional systems and conservation of local natural resources can be one area for investment without much effort. Organic farming of NWFPs can be the first choice for investment (Sharma, 2001).

(ii) ***Eco-tourism:*** Investment in eco-tourism can prove to be a large income generator in the arid lands. Increasing numbers of people are spending increasing amounts of time and money enjoying the unique resources of these arid land environments, *e.g.* spectacular scenic beauty, diverse wildlife population, relatively clean air and other amenities such as large open spaces. Investment in the forestry sector can be in amenity plantations, desert parks, buffer strips, roadside plantations and green belts can be economically as well as environmentally beneficial for the arid zone.

Exploring the potential of arid zone agroforestry as a tool for solving environmental problems.

Conservation of biodiversity. Biodiversity is the essential safety mechanism of any ecosystem and as the arid zone is considered to be fragile in nature the importance of conservation of biodiversity is further increased. Because of biotic interference and developmental activities, not only is the population of various species reduced but also their natural habitats have been destroyed. Thus the easiest way to save these species is *ex-situ* conservation with protection. This is possible through arid zone agroforestry, in which all types of plants *e.g.* trees, shrubs, climbers *etc* can be accommodated, combining conservation with domestication for increased production.

Desertification control. Desertification is the result of exploitation of arid zone ecosystems beyond their carrying capacity. The carrying capacity of the arid lands is inherently low thus they are highly prone to desertification.

Through soil and water conservation and sustainable management of vegetation, the carrying capacity of natural resources can be enhanced. Various arid zone agroforestry models, e.g. wind breaks, sand dune stabilization, silvi- pasture, agri-horticulture etc. can be appropriate for soil and water conservation in arid lands (Gupta and Sharma, 1997), and the productivity of lands suffering form desertification can be enhanced and in this way desertification can be controlled.

Arid lands can be considered simulated desertified drylands for planning of desertification control. This can be important for preventive strategies for drylands rather than to control desertification after the process starts. Any technology or strategy that is found suitable to increase the productivity of arid lands, would give better results if applied to the desertification-prone drylands because they have a better environment than the arid lands.

Assessing the potential of arid zones as carbon sinks. In addition to their other benefits, trees on farms or in pastures also store carbon. Arid lands cover almost one forth of the earth's surface and any effort to increase tree populations substantially contributes to their value as carbon sinks. Due to their perennial nature and high root turnover, grasses can also prove to be a good carbon sink. Silvipasture systems are prevalent in most of the arid zone and when both the tree and grass components are properly managed, an increase in net carbon storage can be achieved. The arid lands are prone to desertification and trees and shrubs can be grown on these lands to prevent further soil degradation and to produce biofuel. These trees also enhance soil organic carbon content and thereby increase soil carbon storage. About 104 million ha land area in the world is affected by strong and extreme soil erosion. Assuming a low rate of increase of soil organic carbon content at 40-60 kg C $ha^{-1}year^{-1}$ the potential of C sequestration in soil is estimated at 0.004-0.006 Pg C $year^{-1}$. There are an additional 930 million ha of salt affected soils in the arid and semi arid regions of the world. Adoption of reclamation measures on these soils could potentially increase above and below ground biomass production and increase soil organic carbon (SOC). Assuming that the rate of SOC increase through adoption of reclamation measures is 200-400 kg C $ha^{-1} year^{-1}$, the potential for C sequestration is 0.186-0.372 Pg C $year^{-1}$ (Izaurralde et.al.2000). Thus rehabilitation of arid lands can contribute to a great extent to solving the green house problem of the world.

Conclusion

Arid zones are known to be fragile ecosystems and interference without knowing the ecological linkages within these systems can result in their degradation, which is often irreversible. Agroforestry as traditionally followed in these regions is a complete, ecologically sustainable livelihood system. To meet both present and future demands, policies need to be supportive of the development of these traditional agroforestry systems based on synergism with nature. There is also a potential to utilize arid zone agroforestry to solve global problems like desertification, global warming (through increased carbon sequestration), and biodiversity conservation, which will require that greater international policy support for arid zone agroforestry.

3

Cold Arid Horticulture: An Overview

Indigenous knowledge in the Himalayan region is the inter-generational wisdom of local inhabitants to perform their livelihood operations in a most eco-friendly manner under remote, isolated and inaccessible conditions; characterized by harsh climate and limited survival options. Since this knowledge is transferred orally from one generation to the next, it is dynamic in dissemination and scientific in indigenous experimentation; receiving constant stimuli from outside. However, indigenous not only stands for ingrained intrinsic knowledge, but is also amenable to modifications based on latest technical know-how by local inhabitants through native means to suit their daily requirements. Therefore, to discard any indigenous knowledge on connotations of superstition, conservatism, primitivism etc. by modem science would only result into a failure of the developmental networks. Evidences have shown that the developmental projects which overtly rejected already acquired knowledge of the local inhabitants, have failed to achieve their targets.

Throughout Himalayan region, watershed resource use and productivity is based on crops, horticulture, pastures and forestry which is largely influenced by geographical and environment diversity prevailing in its different zones.

The wide variations in altitude and other agro-climatic parameters such as rainfall and temperature, broadly classify Himalayan region into four major agro-climatic zones. These include (1) the low hills and valleys near the plains, (2) the middle hills and valleys with sub-humid climate, (3) high mountains and valleys with temperate climate and (4) cold dry desert zone.

In this document, ITK for upper Himalayas comprising of high mountains and valleys with temperate climate and cold dry deserts are described/presented.

Before turning to ITK *per se,* it is pertinent to present a brief glimpse of the two regions under consideration. First, the cold desert region and then the temperate zone of the western Himalayas/Indian Upper Himalayas.

Cold Deserts in Western Himalayas

Spread over an approximate area of 74, 809 sq. km, the cold desert area in India covers 12 out of 131 desert blocks in India. Leh and Kargil districts of Ladakh in Jammu & Kashmir and Lahaul and Spiti along with some parts of Chamba and Kinnaur districts of Himachal Pradesh comprise this cold desert area. Similarity in their physiographic location and the consequent geomorphic processes still unfolding in the region lend to this entire region a largely similar texture.

Located in the interior of continents, away from any source of moisture, cold deserts manifest remarkable ecological variety and biological diversity. Their geographic remoteness and unfriendly climatic conditions greatly constrain economic growth and development. Environmental degradation, which is on increase, is an additional cause for concern.

Bio-physical Features

A rarified atmosphere, fast blowing winds -eroding the immature sandy soils, extreme variations in daily and seasonal temperatures along with scanty or no precipitation during spring and summer ensure short growing seasons (2-5 months) with exposure to harmful infra-red and ultraviolet radiations. Unharvested glacial melts, frozen soil moisture during early spring and low relative humidity during the growing season are some abiotic features. Sparce natural vegetation results from over-exploitation by a variety of agencies - e.g. grazing by both domestic and migratory animals, harvesting (of vegetation) to meet energy needs viz. fuelwood and dry fodder for winter besides demands from pharmaceutical agencies. Additionally, both migratory birds and rodents utilize dispersed seeds as a means of sustenance thereby jeopardizing natural regeneration.

Features such as high transpiration due to excessive heat (often causing mortality), inadequate photo-hours especially during winter, injury due to frost causes poor seed germination, poor plant growth, poor root formation, deformed canopy, reduced radial growth etc. and other physical signs/phenotypic manifestations which in turn affect the productive biomass production in the region.

Natural vegetation is overwhelmingly herbaceous - comprising of a few tree species and a few shrub species. *Juniperus wallichiana, J. communis, Caragana* spp., *Artemisea* spp., *Lonicera* spp., *Potentilla* spp., *Myricaria* spp., *Koleresia dutheii, Ephedra, Salix,* spp., *Juniperus* spp., *Rosa* spp., *Caragana* spp. *Rhododendron* spp., *Betula utilis* are found here. Additionally, manmade forests of poplars, willows, *Hippophae* spp. and *Myricaria* spp. can also be seen along river banks, rivulets and *nallahs.*

The herbaceous element is comprised of *Thymus, Medicago, Trifolium, Anemone, Potentila, Epilobium, Verbena, Allium, Aconitum, Delphenium, Aquilegia, Primula, Geranium, Polygonum* and *Cannabis.* This abundance of the herbaceous element, both in Laddakh and in the cold desert of Himachal, has been the mainstay of the traditional medicinal system prevalent in this region. The nature of flora along with man made interventions have ensured a land use pattern typical to this region.

Socio-economic Features

These features also place constraints on economic growth. Fundamentally these are re-enforced by the biophysical features discussed earlier. Society steeped in religion along with class/case hierarchies, absence of adequate means of livelihood, absence of women's organizations accompanied by rudimentary infrastructure - viz. poor communication network, inadequate developmental institutions/agencies, lack of educational/vocational facilities and inadequate outreach for relevant transfer of technology characterize cold desert society. The result, quite obviously, is that there remains a largely untapped and underdeveloped industrial potential even in the agriculture/horticulture sector that are the mainstay of communities inhabiting cold deserts.

In this region as a whole the initial land use pattern was purely agriculture. It has, however, changed over a period of time to agri-horti-silvi-pastoral. Apple, potato, walnut etc. are some important horticultural crops while poplars and willows have been popularised especially in Lahaul and Spiti and Laddakh to increase both the green cover and also to augment fodder and fuel needs.

The physiographic location of Laddakh enables this district to have only one cropping season - Kharif, which extends from March/ April to October. Further depending upon altitude the growing season varies between two months (above 4000 in amsl) to five months (below 3000 in amsl). Millets followed by wheat are the most important cereal crops. These are followed, in order of importance by fodder crops, barley and pulses. *Alfa-alfa* is the most popular fodder crop. Fruits and vegetables - the horticultural component, are a recent development. Since they yield high economic returns increasingly more and more area is being brought under crops such as apples, raisin grapes and apricots. Mustard, Pea, Lathyrus, millets and turnips are also grown. The central belt (3000-3500 in amsl) is very well suited to the production of vegetable seeds. Rotation in the cropping pattern is hardly practiced.

The fauna of this entire region is quite unique. Due to poor/ rudimentary communication facilities the yak has been the major animal for burden. Besides yaks, livestock comprise mainly sheep and goats.

While the major fauna species of cold desert in Jammu & Kashmir are snow leopard, ibex, snow cock, partridges, magpie etc. A large number of migratory birds visit the lakes and rivers. Pashmina goat, Changthangi sheep, yaks, donkeys and double humped camels are animals of economic importance.

The ibex, bharal, brown bear, tibetan wolf, nayan, marmot, snow leopard, lynx, weasel, vole, snow cock, snow partridge, chukor, chough, raven etc. are found in the cold deserts of Himachal Pradesh. This uniqueness is being preserved through the establishment of two sanctuaries namely the Pin valley National Park in Spin and the Sechu Tuan Nala in Chamba.

As mentioned, yak and sheep dominate livestock composition. Additionally land extensive management and transhumance along with the barter system of converting livestock into other usable commodities are features that characterise animal husbandry practices. Chief among the many problems faced by the livestock sector are - insufficient supply of fodder, overgrazing right up to alpine meadows and difficulty in stall feeding in snow bound areas etc. The *Gaddis* - a migratory tribe of Bharmour, use goat milk to supplement their diet.

Temperate Zone in Western Himalayas

The temperate zone in western Himalayas corresponds to zone III drawn up on agro-ecological basis by the planning commission. It is physiographically a largely mountaneous tract (1800-2000 m amsl).

Bio-physical Features

This zone is characterised by mountaneous tracts of varying altitudes, steep slopes etc. Steeper slopes receive higher intensity of radiation and are not conducive to the growth of vegetation. Conversely, vegetation abounds on the relatively gentler slopes. Moisture from snow-melt and rich organic matter generate dense vegetation of fruit and forest trees. Most of this zone consists of granite and other crystalline rocks of unforseliferous sediments.

An average rainfall of about 100 cm received during the monsoon months and its erratic distribution further necessitate suitable watershed management interventions.

A calendar year is generally divided into three main seasons viz. winter (October-February), summer (March-June) and monsoon (July-September) with a brief spring (mid February-March) and autumn (late September-October). Winter temperatures generally remain below 5°C and precipitation in the form of both rainfall and snow result from the western depression. The cold wave sets the migration of the nomadic shepherds to warmer valleys in the Himalayan foothills.

Summer temperatures remain above 20°C especially during April-June. The relative humidity remains about 40 per cent and the occasional hailstorms are known to cause extensive crop damage especially to apple, plum, apricot and peach. Rainfall during this period amounts to approximately 30 per cent of total annual rainfall.

Monsoon results from the South-west monsoon and about 50 per cent of total annual precipitation is received during this period. The temperatures are known to drop by 2-4°C and the relative humidity reaches 60 per cent. In addition to precipitation the "directional aspect" is crucial for development of vegetation. The Northern aspect receiving less direct sun-shine and facing the snowline has consequently, lower temperatures and high moisture retention which inturn create ideal agro-climatic conditions for the cultivation of temperate crops, especially fruit crops. The southern and western aspects suffer high moisture loss since they receive more direct sunshine and consequently they support a poor vegetation cover. The difference in temperature on the different aspects of hills can also be explained in terms of differential isolation i.e. by proximity or distance from the equator, for example isolation factor for the southern aspect is about 1.5-2.4 times higher than that for the northern aspect.

Socio-economic Features

With horticulture occupying a prime position the land use system can be defined as horti-agri-pastoral. Forests account for about 25 per cent of the total geographic area of this zone. The forest cover varies from thick to sporadic. The chief species are kail, deodar, walnut and oaks etc. varying of course with altitude. The grasslands here are heavily grazed both by draught animals and by migratory graziers. The grass cover comprises mainly of *Themeda, Arunidnella, Hamertheria, Heteropogon* etc. and the legume component is mainly made up of white clover. While traditional millets such as Kongni, Cheenee, Kodda and Bathu, etc. are fast disappearing. Cash crops such as potato and apple form the backbone of the economy with apple alone accounting for more than 78 per cent of the area under fruits. Pome and stone fruits are also grown in this zone. It is estimated that this zone alone contributes 96 per cent of the total temperate fruit production in the state. The traditional almost static farming has changed into a dynamic horticulture led system. Further, with the emergence of the market, farm strategies are no longer focussed only on security and precautionary motives.

Livestock composition is dominated by small ruminants due to availability of larger 'support' area. Cattle form the bulk of livestock rearing and are dependent on natural grass alone. Significantly, fodder is not grown. Hay, collected from natural grasslands is the main source of livestock feed.

The introduction of apple to this zone (in Himachal Pradesh) was the comer stone of the overall horticultural development strategy. Apple production accounts for nearly 40 per cent of the fruit area and about 90 per cent of fruit production. The economic transition was facilitated by the simultaneous development of supporting infrastructure by way of research and extension mechanisms, higher budgetary allocations, government policy initiatives (e.g. launching of the small farmers developmental agency; SFDA etc., an ever increasing focus on infrastructure e.g. roads, market development and Agricultural credit concentration, especially for capital demanding cash crops, in zone of the state is truly impressive.

For decades prior to economic transformation/transition societies in the both temperate and cold desert regions, as elsewhere, were part of economic conditions that centred chiefly around the security and precautionary motives. As such, production of hardy crops e.g. millets etc. was common practice. In the absence of structured markets as we have today, barter trade was resorted to. The absence of physical infrastructure, low economic resource base etc. translated people to adapt their lifestyles to suit their specific locational situations. Stated simply, subsistence farming with a focus on crops that met this goal was the common practice. These 'adaptations' ensured that agriculture was practiced on 'sustainable' lines with better income distribution and equity consequences. Undeniably, such farming/agricultural practices evolved under low population pressure in the different agro-ecological regions and depended entirely on locally available resource, yet, they do hold lessons for us even today.

This is to be appreciated in light of the realization that increasing populations and the consequent reduction of per capita resource availability dictate that a more efficient resource utilization regime be adopted. Such a regime is embodied in indigenous technological practices. Attempts to document such ITK also serve the purpose of preserving such knowledge for posterity, otherwise much of it would be lost due to the largely oral tradition of transferring such knowledge to successive generations. Such an exercise also has relevance for development agencies and extension services for providing relevant (technical) support for sustainable development and management of natural resources.

Soil and Water Management Techniques

In mountain watersheds, irrigation has been practiced as an art for about 3000 years now. Historical records bear testimony to the existence of a number of irrigation works in different parts of the country. In the Himalayas, the perennial river Ganges made it relatively easy to divert its flow through inundation channels. In the south, where rainfall is scanty, the practice of trapping rain water in large tanks and ponds for agricultural purposes is widely adopted.

From time immemorial, surface irrigation methods have been followed. The most effective irrigation method for a particular area depends on the slope of the land, the nature of the soil, the type of the crop and availability of funds.

In mountain areas, water continues to be the scarce commodity not only for irrigation but even for drinking and other domestic uses. This difficulty has been experienced very frequently, inspite of the fact that important rivers namely Sutlez, Beas, Ravi and their tributaries originate from these hills. The existing resources are further declining due to heavy biotic pressure and lack of management of existing resources. Most of our Agricultural/Horticultural activities are carried on under rainfed conditions and this require proper management of available water to be conserved for dry periods.

Sources of Irrigation Water

In the hill region, the scope of boring tubewells, canals and even lift irrigation is limited, such facilities are confined to the low laying areas. Therefore, the most common source of irrigation remains the small water channels locally called Kuhls which intact accounts for 85.83 per cent of the total area under irrigation in hills.

In cold deserts, some villages get water for irrigating their lands from some perennial torrents. In Spiti valley, the source of irrigation water is generally local nallas. Glacial water in cold deserts of Himachal Pradesh which forms the prime source of sustaining life in the region is brought to the field by making Kuhls (Water Channels).

In Kinnaur and other regions, the source of irrigation as well as drinking water is melting snow on the high peaks which runs downward in the shape of small and big nallahas (streams) and also spring out at certain points.

Construction of Kuhls (Water Channels)

In cold deserts of Himachal pradesh kuhls (water channels) are built along the hill gradient for maintaining proper gravity for irrigation. Kuhls are commonly found in West Himalayas cold deserts. The technique for the preparation of kuhls for irrigation purposes seems to have originated since Babylonian times, it is still one of the commonest ways of bringing water to the crops. If the river has a steep gradient, water is diverted into a canal some distance upstream and led along a contour so that it can flow to fields by gravity.

In dry temperate zone, kuhls (wooden water channels) are generally made by making notches at the natural water sources and the water is diverted to the fields for irrigation to different terraces, using the natural gravitational flow of water. Since the topography of the area consists of very high slopes and rocky terrain's, wooden water channels are used at many places as water passes from one place to another. The water channels are built and managed by the villagers with no government assistance. In the lower areas of H.P. bamboo pipes are commonly used as irrigation channels on depressions/ small nala.

In west Himalayan cold deserts for the optimum harnessing of water for irrigation, water channels are constructed along the natural gradients. The irrigation channels (kuhls) are diverted from river tributaries by making use of the natural gradients thus the level of water is higher than that of the cultivated fields.

In upper Kinnaur, the channels (kuhls) are simply dug in the ground to regulate the flow of water. However, where the digging of channels is difficult or the channel has to pass through a village path, underground channels covered with slates are constructed. However, in some parts the wooden channels are also used which are put like a bridge over the path. These channels are made by making a deep grove in the tree trunk or a thick branch.

Distribution of Kuhl Water in Fields

In the cold deserts of Himachal Pradesh, participatory management is employed for distribution of water. All disputes regarding the distribution of water through kuhls (water channels) are amicably settled without hampering the water requirement of any period.

In the West Himalayan cold deserts, all the irrigation channels (Kuhls) cannot be run satisfactorily due to non-availability of sufficient water from Nallas/Khads. This is because of scanty snowfall during the winter months. The majority of hamlets, which lie on the plateaus on the sides of main river get water from the streams which trickles down from the cliffs overhanging the plateaus. These hamlets are the worst off for water, for in the year of scanty snowfall, the streams dwindle quickly and dry up in the beginning of August. Additional snowfall in winter results in less water in natural springs during the season, whereas less snowfall in winter result in the reduction of level in natural springs during summer and consequently crop suffer.

Kuhls are a time tested community made water channels for sharing the glacial water for ensuring cent per cent irrigation in otherwise dry and porous soils.

In Spiti valley, the farmers have developed the irrigation water distribution system on the basis of their land holdings, in which every field is irrigated timely. So there is no dispute regarding the maintenance of kuhls and irrigation water distribution.

In Kinnaur and other regions, nallas passing through a village are harvested on turn basis called pala. Temporary channels are dug by the farmers towards their fields. The whole community is divided on the basis of number of farm families and one family gets one full water day to irrigate their fields turnwise. For example, if there are

20 farm families in a village, the turn falls after every 20 days. But two adjoining families may share the water for half day each when there is turn of either of the two families. This way these two families get a chance to irrigate their fields after a gap of 10 days rather than 20 days. This way the distribution of water is so well managed that maximum use of water takes place in a particular village. The turn of a family comes/starts around 2000 to 2200 hrs on a particular day and all the members of the family are engaged in the job on its turn.

In upper Kinnaur, the irrigation technique is much more pronounced. The fields are generally divided into small compartments by making earth bunds to allow water to stand in the field for a longer duration for saturating the soil. Hence need for second irrigation arises only after 20 to 25 days even in those agricultural crops which otherwise require irrigation after a gap of 10-15 days. At the first turn of irrigation, first compartment is irrigated; followed by second and so on. On the second turn of irrigation, however these compartments are irrigated in reverse order, i.e. sixth compartment is irrigated first followed by fifth and so on.

In temperate areas of cold deserts crop cultivation without irrigation is not possible because precipitation takes place in the form of snowfall. People take advantage of glacial water and perform collective operations for effective distribution and ensured supply of this scarce source. The management of water in a particular field is regulated by apportioning into different compartments because of the season. The mouth of first compartment is closed to regulate the flow of water towards the second compartment. The same method is adopted to irrigate the following compartment. This results in raising the height of channel in front of the first compartment than the channel in front of the second compartment and so on. Now when this field is irrigated during its second turn, the water flows straight towards the fourth compartment. This practice prevents the washing off the upper fertile layers during irrigation.

In the entire Spiti valley, the first irrigation is done 40 days after sowing of crop takes place during April. In the initial stage of watering from the Kuhl to field, the ladies bring water to the field by the use of Urma which is made from animals horn. As per the turn pertaining the Baraghar watering/irrigation is done by constructing small beds in the fields. This method is time consuming and laborious. But on the other hand this method checks the loss of nutrients by leaching. Uniform watering of the plants with equal flow, checks the nutrient loss from field to field and from one bed to another. In Ladakh and other regions standardized irrigation schedule for different crops is followed.

The general schedule is:

Irrigation number	*Stage of crop growth*	
	Local name	*English name*
I	Tol Chu	Germination
II	Sak Chu	Growing
III	Non Chu	Flowering
IV	Gep Chu	Seed setting
V	Do Chu	Crop ready for harvest

The "Gep Chu" or 4th irrigation depends on the colour status of the crop. If crop seems yellow in colour "Gep Chu" is delayed. However, the colour position is blackish "Gep Chu" is hastened.

The farmers have developed irrigation schedule matching the stages of crop growth. Thus, irrigation during critical stages results in the maximization of crop yield as well as water use efficiency.

Use of Kuhl Water for Running Water Mills

Kuhls are built along the hill gradient for maintaining proper gravity for irrigation and running water mills. Wooden water channels are also used for running water-flour mills. These wooden channels are generally made by making notches at the natural water sources and the water is diverted to the water mill, using the natural gravitational flow of water. Since the topography of the area consists of very high slopes and rocky terrain's, wooden water channels are used at many places as water passes from one place to another.

Granite stones are used for grinding food grains. Long wooden channel placed at steep gradient is used for maintaining the high speed of the water flow. This is necessary for maintaining the high speed of the water mills wheel.

Now a days water mills are very rare. Water mill technology is in an extinct stage, because of power supply availability and less grain production. Food is purchased from cooperative societies or private shops now a days.

Methods of Irrigation

Flooding of glacial water for higher crop productivity. In most Himalayan cold deserts water is brought in channels from glacial melts for irrigating the fields. Flooding the fields with the glacial water for improving crop productivity is also common.

The deposition of fresh silt with unweathered minerals (especially lime) forms glacier source of fresh salts. The glacier melted water is often below 2°C which protects the crop from different kinds of diseases.

Indigenous Drip Irrigation

The practice of using pitcher water as a source of irrigation on new fruit plantation in sandy loam/loamy sand soils, in areas of canty rainfall is prevalent in temperate districts of Himachal Pradesh. The pitcher is placed in soil and the new plant is planted close to it.

The pitcher is filled with water during summer months (April-June) and stone/slate lid is placed on the top. The roots draw moisture/ water from pitcher which is turn reduces the mortality. The pitcher once filled, supply sufficient moisture for atleast two weeks and then again it is filled with water.

Bamboo Drip Irrigation System

In this system of irrigation bamboo channels (open) are used for irrigating the fields.This system is common in North-East regions of India. Small holes are made at the internodes of open bamboo channels, from where water gets trickled down in the field. These channels are placed along the natural gradients. In these channels, no uniform head for water trickling is maintained.

Manual Irrigation in Vegetables

In the initial stage of watering vegetables, people bring water to their fields with the help of buckets and in Spiti valley ladies bring water to their fields using Urma which is made from animals horn. In this method after bringing water in buckets, water is supplied to the vegetables with the help of lota (mug), whereas in case of Urma, irrigation is done by constructing small beds in the fields. But this method is too laborious and time consuming.

Water Harvesting Methods

Small Ponds for Spring Water Collection

Another method is the collection of spring water in small reservoirs scattered at intervals on the high uplands and then drawing water from these ponds when required. It is a common practice in cold deserts and temperate wet Himalayas. Water from these ponds is used for irrigating crops and also for drinking purposes.

Harvesting of Dew and Fog Water

In plains and in valleys occurrence of dew and "pale" is very common after the receding of monsoon. After monsoon the humidity remains quite high (85%) in the atmosphere. During night time, temperature falls down sharply resulting in the formation of more water molecules from vaporous. As they are heavier, they fall on soil surface and make the layer moist and wet.

In the hills, there is traditional practice to plough the fields early in the morning before dew or fog water is evaporated. By ploughing, moisture is mixed with soil particles in the plough layer i.e. 9"-12". This moisture is well retained by soil. If soil is clayey in nature, retention of water remains for a longer time and becomes a source of soil moisture. It is quite useful for land preparation in October-November and for the sowing of rabi crops like wheat, barley and pulses.

Roof Water Harvesting

In the lower areas of Himachal Pradesh during the rainy season, roof water is collected in dugout structures which are known as "diggi" in Kangra district and "'Khati" in Hamirpur and Bilaspur districts. These structures are dug in hard rocks. Not only roof water but also surface water is collected in dugout structures.

Harvesting of Rain Water

In the hills, rains are erratic and torrential. Relatively high percentage of rain water goes as run-off and stream flow. It carries fertile soil and plant nutrients which makes the soil degraded and barren. In some areas this excess water is stored directly in the farm ponds, depression or stream flow or is diverted to safer points where it is stored.

The stored water in ponds and depressions is used for irrigational purposes, as a life saver or for supplementary irrigation during lean periods. It is also stored in dugout structures. In some areas during summer, it is used as drinking water humans, livestock and for other domestic purposes.

The ponds with time are sealed, with silt and clay particles thus infiltration/percolation losses are reduced and ponding time and volume of water is increased.

Harvesting of Water from Snow Melting

Harvesting of water is also done by constructing water ponds and water is collected in these ponds from melting snow.

Use of Pang (Spang) Grass for Controlling Seepage and Side Losses in Water Tanks and Irrigation Kuhls.

In Ladakh Pang (Spang) grass is used as the inner lining of zings (water ponds) and irrigation kuhls for checking percolation losses. The use of spang grass which is growing profusely in Ladakh, explains its non-permeability properties similar to that of polythene sheet or cement lining. Its chemistry is required to be analysed, as the farmers claim its utility in water retention is far superior than the polythene/cement.

Moisture Conservation Through Mulching

In Kinnaur, covering the surface of soil with chilgoza tree needles and grass from the Kandas (hill tops) is a common mulching practice. Mulching conserves soil moisture in the fields. It also helps in the moderation of soil temperature. In this way hydro-thermal regime of soil is improved. However, the continuous use of chilgoza tree needles increases the acidity of the soil.

In the hilly areas, ploughing is done, which aids in moisture conservation, as the soil acts as mulch.

In Ladakh, farmers regulate optimum irrigation by inserting a belcha (spade) in the soil If it is completely inserted (front portion), the land is considered to be properly irrigated. Similarly, in a few other cases, mud is thrown in the air. Its splitting into pieces shows proper irrigation. Complete insertion of the. front portion of belcha (spade) or throwing of mud in the air and its consequent splitting into pieces indicate the soil moisture level at field capacity, where 100 per cent moisture is available to the crops.

Drainage

During rainy season the rains are torrential, which causes splash erosion resulting in the sorting of particles and the formation of false compact layer on the surface. It yields water pounding and subsequently water logging. Crops such as maize, capsicum, tomato which are grown during this season are very sensitive to water logging. In our traditional agriculture there is a common practice that during the preparation of a field the slope of a field, is kept inside which is provided with a channel to take excess water from that field to a safer place, from where it is disposed to stream or nalla through grassed water ways. The grassed water ways are kept permanently and help in the drainage. These channels and grassed water ways are positioned in such a way that they do not hinder any agricultural activity such as ploughing, hoeing and harvesting.

Use of Smoke for Protecting Fruit Crops from Frost Damage

In the lower areas of Himachal Pradesh, Mango plants are mostly damaged by frost injury during winter months i.e. December and January. Smoke layer protects the mango plants from frost injury. This practice is common in the lower areas of Himachal Pradesh.

Soil Management

Cultural Practices

In West Himalayan region, in the month of March/April, when snow melts and weather condition improves, the bunds and comers of the fields are dug-out and weeds and grasses are removed with the help of spade and clods. The grasses or weeds are beaten up and then soil is separated from these clods and collected in lower fields.

This practice of removing weeds and grasses from bunds and corners by digging helps in weed control in the cultivated fields. Secondly area under crops remains the same as that of previous crop i.e. area is not wasted for weeds and grasses. Thirdly the soil added in lower fields from the bunds of upper field is rich in nutrients and it improves the soil fertility.

Use of Broader Plough in Upper Valleys

Ploughs are broadened in Ladakh by attaching flat wooden pieces to both sides of the iron blade. This indigenous plough is preferred over the one available in the market.

This technology seems to have twofold functions of saving labour and that of stabilizing the loose sandy strata in one ploughing action, which suits the small terraces.

Sheet Erosion Control

It is not a damaging form of erosion, mainly because it is often not recognised and seldom treated. It accounts for the loss of billions of tonnes of soil every year. Due to splash of rain drops particles are knocked loose and then carried away by the runoff. The sheet erosion result into rill and gullies which are controlled by very cheap treatments. Sheet erosion is more apparent in forest areas that are devoid of ground cover or wastelands with very few standing trees.

There is traditional practice to keep surface maximum covered with grasses, shrubs etc., grazing is done in rotation and is allowed only during certain times. It is avoided during the flowering and seed setting stages of grasses. Fibrous rooted shrubs and grasses planted as hedges along the contour of the land slow the runoff, weaken the erosive power of water and cause it to deposit its load of valuable soil behind the hedgerows. As a result the runoff proceeds gently down the slope where hedges have been planted at the correct vertical interval without erosive effect In the foot hills, erosive capacity of stream flow is also reduced by spurs of loose boulders.

Traditional Rainfed Farming

In the hilly areas, most of the area is rainfed except for a few pockets in valleys where irrigation facilities are existing. The choice of crop and rotation, completely depends on crops which require less volume of water. For rotation, legumes are important as mixed crop. During rotation, when rainy season erosion permitting crops are grown, such as cowpea, 'kuth', these form integral part of the mixed cropping system. The crops are chosen as per their nutrition e.g. from old ages protein rich pulses are part of cropping pattern. The coarse grains like 'phaphra', 'chulai' are also grown very commonly which are very rich in nutrition.

Within the premises of the house it is mandatory to have fruit plants such as citrus, mango, anar which provide seasonal fruits rich in vitamin C., carbohydrates etc. The fields are well protected with biofence of thorny shrubs or their cut pieces.

The traditional rainfed farming is done irrespective of land with respect to slope and other characteristics. There are chances of sheet erosion but with traditional knowledge, crop rotation is adopted in such a way that during peak runoff periods sowing of close growing crops provide protection to the soil.

Terracing

From old times, land in the hills has been put under cultivation on scientific lines as cultivation is done up to 25-100 degree slope, where there are many chances of landslips, sheet and gully erosion. But with 'bench terracing practices' the menace of soil erosion is controlled and is very common in hill fanning.

The terraces are constructed across the slope i.e. along the contour. The size of the terrace is decided by the prevailing degree of slope. The terraces are supported by risers of suitable heights and width. The height of riser is again decided by the degree of slope. The risers are sometime made of loose boulders supported by grasses. The roots of grasses help in binding and keeping the boulders intact at a place. The roots of grasses help in drainage of excess water. With the traditional knowledge, farmers are keeping the risers toward inner slopes. In paddy growing areas the risers are erected to facilitate the pounding of water in the field. This type of bunding and terracing is continuing from centuries and terraces are still intact. The bunds are again used for growing palatable grasses which is used as fodder for livestock and trees are meant for fuel, fodder and fibre. The examples are beul, shisham, mango etc. In lower areas the bench terraces are known as "khet". Sometime on the risers contour hedge of grass like khus, local grasses also established.

Use of Maddim (A Plain Wooden Structure) for Field Levelling

Maddim is used for levelling ploughed lands. A heavy stone is put on the maddim for increasing the pressure required for levelling. Sometimes, a man may also sit instead of a heavy stone.

Such an indigenous technology for field levelling is called planking. With this practice, there is very good seed soil contact and very good germination of the crops. Secondly, there is moisture conservation in the fields. Thirdly small soil clods are pressed and broken into finer particles and this way soil structure is improved.

Curved land ploughing for intensive land preparation; soil conservation and water retention.

In west Himalayan cold deserts, ploughing is done in a curved (sword like) manner from the bottom to the top of the slopy land holdings.

Ploughing land holdings in a sword like pattern ensures proper land preparation which includes proper ploughing of the corners which otherwise would have remained unploughed. The ploughing of slopy lands from bottom to top also helps in soil conservation as it checks the loosened soil strata falling from the upper side to the lower. The curved pattern is useful in maintaining infiltration rate of water which otherwise gets wasted with sudden runoff.

Conserving Productive Soil Layer Against Wind Erosion

In west Himalayan cold deserts, fields are irrigated in autumn so that the top layer is prevented from being blown away. In spring the moistened soil eases ploughing.

The productive soil layer, which is very thin, needs conservation against heavy wind erosion, a common feature of the cold deserts. This appropriate soil conservation technique also helps in easy and timely ploughing for meeting the requirement of short growing season.

The moist upper layer of soil which gets frozen in winter also serves as a protection against wind erosion.

Cultivation of Levelled/Flat Lands for Preventing Soil Erosion

In west Himalayan cold deserts, cultivation practices are confined to the levelled/flat lands only.

This practice helps not only in the rational land use but also checks soil erosion in otherwise sandy and loose strata.

Contouring of slopy lands: Ethno-engineering for soil conservation

In west Himalayan cold deserts farmers have developed this technology for cultivation of slopy lands by constructing terraces comprising of plots and sub-plots by using small stones. Stone wall fencing is also constructed for individual land holdings. Terracing of slopy lands helps in conserving soil and moisture and prevents soil erosion. This also helps to carry out other field operations including proper use of irrigation water for checking the surface runoff.

Use of Loose Boulders Spurs for Reducing Soil Erosion

In some areas, people use loose boulders spurs for reducing *the* cutting effect of stream flow in a small nalla (Choes). Use of loose boulders diversion dam with spillway in centre for reducing soil erosion. This method is very common in lower areas of Himachal Pradesh. These dams are constructed across the streams for controlling soil loss.

Use of vegetation, live check of bamboo pieces and loose boulders.

In this method of vegetation, live check of bambo pieces and loose boulders are used for controlling gully erosion. This practice is also common in lower areas.

a) Vegetation
b) Live check
c) Loose boulders stabilized with grasses.

Use of River Bed Soils

The river bed soils are used for raising crops. These soils are rich in nutrients as nutrients are removed with soil from hilly slopes and are deposited in the river beds.

Soil Fertility Management

Proper soil management, ensuring continued maintenance and building up of fertility at a high level is indispensable for the profitable use of agricultural lands. While chemical fertilizers introduce extra concentrated supplies of readily available plant nutrients to the soil, the beneficial effect of organic manures predominantly lies in furnishing humus forming material to bring about improvement in the soil structure, water holding capacity, microbial population and its activity, base exchange capacity and resistance to soil erosion. Much of the plant food removed by the crops is restored to the soil through the application of organic manures.

Soil Management by Crop Residue Harvesting

This practice is prevalent in west Himalayan cold deserts. Barley and wheat stumps (in Zanskar) are pulled out by hand along with the complete root system. Soil is softened by a light irrigation a day before. Wheat is often pulled out while standing, but kneeling or squatting is practiced for barley. Handful of these plants are beaten up against the legs (occasionally a small apron is worn) to shake off most of the earth. These bundles are then piled up like the tiles of a roof. The ears of the lower row are covered and protected from birds by the roots of the upper stacks. In cold deserts of Himachal Pradesh, barley and buckwheat (in double cropping farming system) are also pulled out by roots. This helps in uprooting weeds, soil loosening and porosity maintenance for the coming crop.

Other practice is to harvest crops as close to the grounds as possible. The roots are made to stay in soil for humus production. Very little plant material (stem and roots) is allowed to be left in the soil (in Ladakh) as a protective measure against the soil borne diseases. This practice also increases the fodder resource in winters. Retention" of roots in soil (in single cropping) contributes towards humus availability which improves the soil structure, porosity and water holding capacity of the soil.

Soil Mixing with Night Soils

This practice is prevalent in Ladakh and other parts. Soil with human excreta is mixed and broadcasted over the fields during winter months. Soil is collected from cultivated land holdings and particularly from field bunds of sub-plots for mixing.

The night soil/human excreta possess immense manurial potentiality as it contains the major plant nutrients like nitrogen, phosphorus and potassium. So the addition of night soil/human excreta along with soil from cultivated field improves the soil fertility. The practice of collecting soil from cultivated land and fields helps in easy ploughing during summer cropping.

Organic Manuring, Collection and Management

Organic manures derived from plant and animal resource, are valuable byproducts of farming and allied industries. Organic manures which is bulky in nature but supply the plant nutrients in small quantities are termed as bulky organic manures e.g. farm yard manure, rural and town compost, night soil, green manure etc., whereas those containing higher percentage of major plant nutrients like nitrogen, phosphorus and potash are known as concentrated organic manures e.g. oil cakes, goat manure, sheep and poultry manure, blood and meat-meals, etc.

Flocks of sheep and goats, contribute towards tribal economy by way of milk, meat, wool and manure. These flocks when taken for grazing are tied with small bags which cover their anal parts so that the excreta falls right into the bag.

This region is highly sandy with low soil fertility status. The collection of dropping of sheep and goats by tieing bags is indicative of indigenous wisdom to meet out the shortage of manure. This manure of the droppings of sheep and goats contains 3% nitrogen, 1% phosphorus and 2% potassium.

In Spiti valley, organic manuring is done once a year because of mono-cropping pattern in the months of September-October after the crop. The manure is broadcasted in the entire field, which is followed by ploughing for thorough mixing. The richest manure is called Chaksa which comprises of human excreta and is collected in separate dry latrine pit. The main reason for its nutritional value is that even the bones of animals are thrown in the excreta which adds phosphorus and calcium to the manure.

The daily per capita availability of night soil, human urine and nutrients contained in it is as under:

Particulars	Faeces (g)	Urine(s)
Quantity (natural condition)	133.00	1200.00
Quantity (dry)	30.30	64.00
Nitrogen	2.10	12.10
Phosphorus	1.64	1.80
Potassium	0.73	2.22

This data shows that night soil and human urine have a great manurial potential with regard to nitrogen, phosphorus and potassium. Due to this potential, it is considered good manure by the farmers.

Secondly cattle dung is collected in heaps within cattlesheds during winter months, so that it decomposes under relatively high temperature conditions. Then it is placed out in the open during summer in the form of heaps for further decomposition. Actually the cattle dung contains 0.2% nitrogen, 0.1% phosphorus and 0.15% potassium and cattle urine contains 0.6% nitrogen, 0.1% phosphorus and 0.5% potassium. Due to these immense manurial potentialities of cattle dung and urine, the use of this manure is very much popular among farmers.

Manuring is required for wheat, paddy and maize, which are the main crops of the Bharmour and Pangi regions. The traditional means of manure are as follows:

i) Dung of livestock, mostly cattle, collected from the sheds, pens and camps of livestock

ii) The leaves and grasses which were used as bedding for the animals, got soaked with the excreta/urine of livestock and were then collected periodically.

iii) Feeding of sheep and goats in the fields: This method of manuring is very much in vogue in those places which are visited by the Gaddi graziers, whether enroute to their camps or on move with their herds. The Gaddis are paid for this benefit. These traditional practices continues unchanged. The only improvement that has been made is that the heaps of cow dung are well covered with something or the other in order to protect them from rains and snow.

iv) In the wet temperate Himalayas, green and dried pine needles are collected in heaps and used as bedding material. Before using as bedding material these pine needles are cut into small pieces.

In the absence of chemical fertilizers, organic manuring is the chief mode of soil fertilization. All efforts are made to collect and use animal dropping and for their subsequent decomposition along with the leaves and grasses which are used in manuring the crops. This is the traditional organic manure and is most readily available to the farmers.

It is the product of decomposition of the liquid and solid excreta of livestock, stored in the sheds, pens and camps of livestock along with varying amounts of straws or other litter used as bedding. This farm yard manure/compost prepared from farm litter, liquid and solid excreta of livestock contains 0.5% nitrogen, 0.2% phosphorus and 0.5% potassium.

To enhance the productivity, people in Kinnaur still use the farm yard manure. It is worth mentioning that here animals are kept primarily to meet the need of manure.

Donkeys, cows, goats and sheep are the main source of manure. The manure is collected either from the cowsheds inside the house or the cowsheds outside the house. Generally, the ground floor in each house is used as a cowshed so that animals can be looked after in a better way during winter months. The dung is put outside the house in a heap form in lower areas, whereas, in upper areas, it is directly put in small heaps in the fields. These small heaps of dung are covered with a thin layer of soil to avoid the dispersion of manure by wind. The manure is directly mixed with the soil while ploughing. Farm yard manure is transported to the fields in Kilta (bamboo container) by people's participation and also by horses.

Amongst the manures, the cowdung is preferred the most. According to most farmers the sheep and goat dung may lead to burning of crops if applied in excess. Ass dung though used is not preferred much. On an average 125 to 250 qtls of manure is used per acre by the farmers throughout the Kinnaur region.

The practice of keeping small heaps of manure in open field with soil coverage in high altitude zones helps in better decomposition due to the maintenance of better temperature conditions. Use of sheep and goat manure in large quantities leads to burning of crops. The burning of crops is due to the toxic effects of high levels of nitrogen, phosphorus and potassium in goat and sheep manures. The goat and sheep manure contains 3 % nitrogen, 1 % phosphorus and 2 % potassium.

Use of Ash in Ladakh

Nutrient Recycling

The inhabitants of this entire region use cattle dung, shrubs and bushes as the main source of fuel. Ashes available, there upon, are mixed either with household waste or human excreta. Sometimes ashes are also broadcasted in the fields. Mixing of ash with household waste and human excreta aids in nutrient availability and recycling. Ash primarily meets the deficiency of potash. Availability of phosphorus is also ensured. In addition to this, human excreta and household waste also contains good amounts of nitrogen, phosphorus and potassium.

Softening of Hard Soils

In Nubra valley, hard soils are softened by putting ash obtained from cowdung, sheep/goat manure, fuelwood etc.

Through this practice upper layers of soils are not only softened but their fertility status is also improved, as ash contains phosphorus.

Increased size of potatoes through the use of ash and goat manure. A mixture of kitchen ash and goat manure is used in kitchen gardens (Nubra valley) for growing potatoes.

The spreading of this mixture as an organic manure, increases the size of potatoes on account of optimum supply of nutrients in otherwise nutrient deficient soils. Secondly organic manure improves the soil structure, porosity and water holding capacity of the soils. In this way there is an overall improvement in physical, chemical and biological properties like microbial population etc., which has increased the size of potatoes.

Poultry Manure and Ash for Increased Vegetable Production

This specific technology is used only in case of tomato, brinjal, capsicum and cauliflower. Kitchen ash and poultry manure mix enhances vegetable production levels.

Stage of Farm Yard Manure in Cultivated Fields

In west Himalayan cold deserts, FYM with a thin coverage of soil is kept in small heaps in the fields from October to March. With the onset of summer months it is spread in the open field.

Coverage of organic manure (FYM) with soil in open fields throughout winter helps in regulating (heap) temperature necessary for proper decomposition of FYM.

Green Manuring

In Bharmour area the practice of green manuring is localized in a few villages (paddy growing). Leaves and twigs of wild bushes such as basuti and kaimal are used.

Use of Goat Manure

In Ladakh, goat manure is considered to be more nutritious. Goat manure when added to millet fields improves production. Goats are specially penned in these plots/fields.

Goat manure improves not only the millet production but also its taste. According to farmers vegetables grown in goat manure have longer keeping quality. It is easy to plough fields manured with goat excreta. Actually with the addition of goat excreta, there is improvement in the physical properties like soil structure, water holding capacity and porosity. There is also an improvement in soil fertility as it contains 3% nitrogen, 1% phosphorus and 2% potassium.

Use of Sachik Soil for Higher Crop Yield

Yellow soil (Sachik) found in Tagloom area is used as manure for enhancing crop production. Yak loads of this yellowish/dark brown coloured soil are scattered in the fields.

Biofencing with Seabuckthorn (Hippophae Rhamnoides)

This practice is prevalent in Spiti and other regions. There is a common practice to provide biofencing with seabuckthorn in cold deserts in general and Spiti in particular. The biofence of seabuckthorn being thorny in nature protects crop from stray animals. Its multipurpose utility as a nitrogen fixer, checks against soil erosion, conservation of soil and moisture, source of fuelwood and indigenous drug (rich source of vitamin C) makes it a promising plants for eco-economic rehabilitation of the region.

Sprawling of Ash Dust in Cucurbits and other Vegetable Crops

In the west Himalayan cold deserts, ash dust is a product obtained after the combustion of fuelwood. It has been observed that dusting of material in the fields enhance early maturity and high yield of vegetable crops.

The reason for the early maturity of cucurbits and vegetable crops is due to the fact that ash dust contains sufficient quantity of phosphorus in available form to the plants. Secondly, in cucurbits the ash dust has been used to repel the insect pest of the crops. Thirdly, amendments of ash dust in the soil, improves soil structure and fertility. Ash dust is also useful in enhancing the maturity of bulb crops which normally takes 6-7 months for obtaining economic yield.

Drought Power According to Soil Texture

In west Himalayan cold deserts, ploughing is generally carried out by dzos, however in sandy situations horses are employed for its speedy completion. In Turpuk of Nubra valley ploughing is done by a single horse.

Sandy soil have less soil strength than clayey soil. Due to this reason, the drought power requirement for ploughing varies according to soil texture.

4

Participatory Irrigation Management

Productivity and Sustainability of Irrigation Systems

Participatory Irrigation Management (PIM) refers to the involvement of irrigation users in all aspects and all levels of irrigation management.

"All aspects" includes the initial planning and design of new irrigation projects or improvements, as well as the construction, supervision, and financing, decision rules, operation, maintenance, monitoring, and evaluation of the system.

Problems of Irrigation

In a national seminar on PIM in India held in 1994, the participants identified the following problems, or issues, with irrigation management. While one can debate whether some of these are "problems" or "symptoms" of problems, they are certainly issues familiar to all of us:

- Inadequate water availability at the lowest outlets
- Poor condition/maintenance of the system
- Lack of measuring devices and control structures
- Inadequate allocation for O&M
- Inequitable distribution of water
- Lack of incentives for saving water
- Poor drainage.

In particular countries and particular irrigation systems, some problems may be more critical than others. There are always additional problems or issues that could be added.

Participation Improve Irrigation

Now let us consider the problems with irrigation management and think about whether and how participatory management approaches can help solve them. We will consider them one by one.

- Inadequate water availability at the lowest outlets
- Poor condition/maintenance of the system
- Lack of measuring devices and control structures
- Inadequate allocation for O&M
- Inequitable distribution of water
- Lack of incentives for saving water
- Poor drainage.

This new perspective of irrigation management has been applied to both projects as well as policies and has greatly expanded our understanding of why some irrigation systems work better than others. It is not only a physical problem; irrigation also involves a much broader set of management issues. For example, how is the system financed? What type of monitoring is carried out and how is this fed back into the O&M process?

Indeed, if we do a straight line projection of what is happening, we would perhaps predict that the state will eventually disappear entirely from the irrigation over the next several decades. But no one is realistically predicting the demise of the state's role in managing irrigation; there will continue to be an essential management role for the state. What is happening is a rationalization of the respective roles of government and users. The pendulum that had shifted far towards the direction of strong government involvement in all aspects of irrigation management is now swinging the other way, towards a more sensible equilibrium.

Inadequate Water Availability at the Lowest Outlets

If the water availability is inadequate because there is inadequate supply, then there are clear limits within which to suggest improvements. But even here, participation in the sense of transparency and dissemination of information about the water situation (e.g., through a system-wide organization) can engender good will from farmers vis a vis the system managers, thus minimizing the likelihood of intentionally breaking gates to increase water supply.

When the users understand the constraints faced by management, they are far more likely to cooperate. If the water availability at the lowest outlets is inadequate because of water thefts upstream, then part of the solution could entail bringing upstream and downstream users into a single management entity that would be accountable to all the users.

This principle of including as many stakeholders as possible in a single organization underlies the very large water user organizations found in Mexico, for example. Participation of the users in managing the system is not the only possible solution, of course.

More diligent management by government irrigation engineers could also solve the problem. But what are the incentives to an government engineer who will be paid by his agency at the same (usually low) rate regardless of how diligently he supervises water use between the upstream and downstream users.

Compare this situation with that of an irrigation engineer hired by a water user organization to oversee water distribution between upstream and downstream users. If he does not satisfy both sets of farmers, he will lose his job, and, at least in the case of Mexico, his salary may be double that of his agency counterpart. This engineer has a very strong incentive to distribute water equitably throughout the system, so that he remains in good favour in the eyes of all farmers.

Poor Condition/Maintenance of the System

The almost universal problem of deferred maintenance resulting in deteriorating infrastructure is on everyone's list of irrigation management concerns. Can participation help? When the users of the system are also in charge of maintenance, they have every incentive to perform timely repairs, to monitor the quality of the work, and to protect structures from vandalism.

When the users also participate in initial design decisions, with the understanding that they will inherit the system as their own, they have a strong incentive to demand a system whose operation poses no technical problems, to monitor construction quality, and if they will repay some of the capital costs, to minimize the overall costs of construction. Most often this translates into reduced demand for lining canals except where it makes good operational sense.

The conservative tendency on lining which participation encourages may in itself result in substantial cost savings initially, and long-term improvements in maintenance (since users are reluctant to repair lining that they do not consider essential).

Lack of Measuring Devices and Control Structures

When irrigation managers are engineers employed by the government irrigation agency, there is little incentive to measure actual water flows. The philosophy of "no news is good news" is the standard approach to measurement. Similarly, control structures require both operational attention and maintenance which may be beyond the interest and/or budgets of the managing engineer. Would this situation change if farmers were in charge of management?

Imagine a situation in which the managing engineer would have to prepare weekly reports to a governing board of farmers, to whom the engineer owes his job. Would he now have an incentive to know how much water is going into each outlet? Would he now have an incentive to control the water so that he can adhere to his operational plan? This type of performance-based management by which the governing board holds the engineer accountable for water flows is theoretically possible under a government administered system. But if we can imagine two governing boards of an irrigation system, the first comprised of elected farmers who themselves depend of that irrigation water, and the second comprised of civil servants who enjoy job security, which of these boards is more likely to demand high performance? Which of these boards would be more likely to be interested in the flow rates in various reaches of the system?

Inadequate Allocation for O&M

Inadequate allocation implies either high costs, or low recovery, or both. The costs of both operating and maintaing the system are increased when farmers deliberately intervene to disrupt the scheduled operation, or to damage structures, and costs are reduced when farmers respect the operating plan and report maintenance problems promptly. Cost recovery appears to be enhanced when farmers can see that the money is going into system O&M (see Vicious Cycle of Public Irrigation), and when they perceive that they are receiving valuable services from the system, i.e., the timely provision of water. Can participatory irrigation management contribute to lower costs and higher recovery? When farmers are directly involved in the design of the system itself and the placement of structures and when they have input into establishing the operational plan, they will be more likely to both understand and protect the facilities and the plan.

When farmers are clearly the owners of the physical system, so that the maintenance costs are their own responsibility, they will have a strong incentive to protect the physical integrity of the system to reduce their overall costs. The cost recovery issue is complex. When farmers are involved in management they have a better understanding of how the funds are being used, and this knowledge is likely to render them more willing to pay the fees. To the extent that irrigation management is improved with user involvement, the farmers may feel more willing to pay for improved service (although these links are difficult to prove). Perhaps the best evidence for a relationship between participatory management and cost recovery comes from actual experience. In the Philippines, the payment rate for irrigation service fees is significantly higher in those systems where water user associations have been formed. In Mexico, payment rates are close to 100% because the system managers — controlled by a board of irrigation users — will refuse to give water without prior payment. Political realities would not allow this type of ultimatum were the managers part of the Irrigation Department.

Inequitable Distribution of Water

Disparities in the supply of water available to farmers are not only due to the physical problems of reaching the tail end of large systems. Disparities may also be caused by intentional tampering with the system, either by physically enlarging outlets, or offering bribes to the water masters. The result is that influential farmers can take more than their due share, thus depriving others who have the misfortune of being located downstream. Can management participation by the water users help solve this well-known situation? Or would the withdrawal of state authority actually encourage powerful farmers to take even more water, with no fear of government sanctions? Social theory suggests that under user management, tampering with water distribution becomes much more difficult. There is greater transparency and accountability rendering illicit water deliveries more difficult to conceal and even more difficult to defend. Whereas thefts of "government" water might be tolerated by other farmers, when the water supply is allocated to the collectivity of farmers, any theft of water implies stealing from a fellow farmer, and a fellow member of the association.

Lack of Incentives for Saving Water

What are the incentives that can encourage farmers to conserve water? Can such incentives become operative more readily under user management or under government management? The set of factors that determine water conservation behaviour are complex. One factor is security of supply. If farmers know when they expect to irrigate next time, they may be more modest in applying water, whereas if they are afraid they might have to wait a long time for the next irrigation, they will be apt to over-irrigated as insurance against this uncertainty. Another factor may be information and management confidence. If farmers are aware that there is an overall scarcity of supply and they believe the scarce water will be equitably distributed, they will may be more likely to limit their own consumption, or at least refrain from taking more than their share. A third factor is the transferability of water rights. If farmers in one part of the command area do not need water this week, but may need extra water after three weeks, can they forgo water now and use it three weeks from now?

While these three types of incentives — security of supply, information and management confidence, and transferability of water rights — can be found under both government management and user management, they are more likely to be found under user management. The reasons have to do with accountability. When the manager are accountable to the users, those managers have a direct incentive to keep supplies regular, and to inform the users of any problems. The continued employment of the managers depends upon the management confidence of the users. Similarly, the transfer of water rights from one user to another, or from one association to another, is not impossible under government management, but it is unusual and perhaps non-existent. Water markets are the formal manifestation of transferring water rights. While water markets are very much a novelty in the irrigation sector, there is a promising potential for tapping market forces to provide economic incentives for saving water, and for using it in the most productive manner.

Poor Drainage

Investments in drainage are notoriously under-valued in government managed irrigation systems. The emphasis is nearly always on increasing the supply of irrigation water, rather than on facilitating disposal of drainage water. It would be unrealistic to claim that farmers are necessarily more foresighted than government agencies in giving attention to drainage, but there is evidence from Mexico that more attention is given to drainage after transfer to user management. Part of the reason in the Mexican case is that the user-managers are able to operate the irrigation system at an overall lower cost, and therefore have funds available for drainage O&M.

Another part of the reason is that the government agency continues to subsidize drainage through monitoring soil and water quality and helping the new managers address priority problems strategically. In Mexico the water user organizations are responsible for both irrigation and drainage. In the Chambal project in Rajasthan, India, separate drainage committees are being established to operate and maintain new investments (by the government) in on-farm drains.

Rationale for Participation

Why participation? Another question might also be asked: "Why should the government be involved in irrigation?" Clearly, there are investments that only the government can make, or where the government has a definite advantage vis-a-vis farmers, even very well organized associations of farmers. Construction of dams and barrages, for example, or large canals, would be extremely difficult for farmers to handle. Governments provide us with available institutional resources — departments, agencies, trained staff, etc. — which can be used to get things done. Why re-invent the wheel and ask farmers to organize their own arrangements for building as dam?

- Comparative advantages
- Improved design, construction, and O&M.
- Lower costs to government.
- Social capital.

Comparative Advantage

Farmers have some comparative advantages as well. They have direct incentives to manage irrigation water in a productive and sustainable manner; they offer an on-the-ground presence that even the most dedicated off-site agency staff cannot equal, and they have an intimate knowledge about their fellow irrigators. The logic of the PIM approach is that both governments and farmers have separate comparative advantages. At the moment, governments are trying to do much more than they can do well. What are the advantages that management by farmers — by the users — can offer?

Improved Design, Construction and O&M

When farmers are directly involved in the design process, whether for new systems or rehabilitation of old ones, they will provide useful design input and they will come away with an understanding of the design logic of the system they will be managing.

During construction, farmer input has the functions of quality control (ensuring design standards are met), cost savings (through guarding against needless spending, and substituting some costs with farmers' own labour), and construction knowledge. Knowing how the system is constructed will help in repairs later on. The advantage of farmer inputs into O&M, either as direct managers or as the overseers of technical managers, has been discussed.

Lower Costs to Government

Cost savings to the government irrigation agency is often the driving force behind irrigation policy reforms. Government run systems are chronically short of maintenance funds leading to deteriorating systems and more difficult operation. Management transfer of major levels of the system to users offers government agencies an escape from this vicious cycle. While some critics see this as merely passing the costs on to farmers, the picture is not usually so bleak. Evidence from Mexico and Turkey suggest that farmers can manage better and more cheaply than their government predecessors. Thus, both farmers and government can benefit from these cost savings; farmers can enjoy better service, and cost savings; the government incurs less management cost and can then afford to improve service in the main system.

Social Capital

The organizations that farmers establish for managing their irrigation systems constitute a form of social capital that can have spin-off effects in other aspects of social and economic life. The network of contacts among agency staff and the water user organization leadership, for example, can bring the farming community into closer touch with related services, e.g., credit, educations opportunities, or even political access. And the skills that farmers learn through their experience with their water user organization — accounting, budgeting, planning, organizing — constitute a set of knowledge that can be used in many other productive endeavours.

What is PIM?

- PIM means "Participation" — not only in O&M and financing, but in making decisions that will affect the O&M — and financing
- Continuum of management approaches and levels of participation
- The concept of users' organization and transfer of management to users is different from PIM.

PIM Means Participation

Participation refers to a continuum of involvement in management decisions. One meaning of "PIM" may be that the irrigation users have total control and responsibility over the operations and maintenance of part or all of the irrigation system. Another meaning of PIM may be that a farmer council plays an advisory role, with real power remaining in the hands of the irrigation agency. Various levels of participation are outlined below.

1. Information sharing
   - Translation into local languages and dissemination of written material using various media
   - Informational presentations and public meetings.
2. Consultations
   - Meetings
   - Field visits and interviews.
3. Joint Assessments
   - Participatory assessments and evaluations
   - Beneficiary assessments.
4. Shared decisionmaking
   - Participatory planning
   - Workshops and seminars to determine positions, priorities, roles
   - Meetings to resolve conflicts, seek agreements, engender ownership
   - Public reviews of draft documents.
5. Collaboration
   - Formation of joint agency/stakeholder committees/task forces
   - Joint work with user groups, NGOs, or other stakeholder groups
   - Stakeholder groups given principal responsibility for implementation.
6. Empowerment
   - Capacity building of stakeholder organizations
   - Hand-over and self-management by stakeholders
   - Support for new, spontaneous initiatives by stakeholders.

Continuum of Involvement in Management Decisions

We can characterize the range of state-user relationships as a continuum from the state doing everything on behalf of the users, to the case of the state doing nothing for the users, other than leaving them alone. In between these two ends of the continuum is a very large gray (or blue) area where a government agency performs some management functions and farmers perform other functions. For purposes of discussion, we can divide the continuum into four types from more to less government involvement:

Type 1: Government does everything. In Malaysia, the Department of Irrigation and Drainage provides for the operation and maintenance of the main and secondary canals, while government sponsored farmers' organizations are responsible for providing water to individual farms. Farmers have no responsibility, and make no management decisions, about the water upstream from their outlets.

Type 2: State dominates; users help. The conventional management division in large irrigation systems is that the state takes responsibility for operation and maintenance of the head works such as a dam or river diversion, and the main, secondary, and larger tertiary canals, while farmers are responsible for managing water distribution and maintenance along the lowest level canals. Typically this entails farmer groups of between 10 and 50 farm families who are expected to work out sharing arrangements on their own.

Type 3: Users dominate; state facilitates. In some countries, associations of water users enter into contractual agreements with state water agencies for the provision of specific water services. In the case of Mexico, the National Water Commission manages the head works and main canals, while legally recognized water user associations employ their own technical staff for the management of the secondary and tertiary levels of the canal networks. Farmers pay their associations for the water, and a small portion of that fee is passed on to the National Water Commission for their services.

Type 4: Farmers do everything: In the Hill regions of Nepal most of the irrigated area is in the hands of local communities who have constructed their own canal systems, generally tapping small stream flows. Similar examples of local, farmer-managed systems can be found in nearly every country where irrigation is important, and the rules and customs of such systems provides a valuable pool of local knowledge that can be tapped in developing new irrigated areas.

Users' Organizations Different than PIM

Management approaches in irrigation generally fall into three categories: (1) public sector management, (2) private sector management, and (3) users' organizations. This last type can be termed "userism," and the process of transferring management to users can be termed "userization". The concept of "userism" is quite different from "privatization" in that we are talking about transferring management not to a third party "owner" who would purchase the irrigation system from the government and then hire out irrigation services to farmers. Rather, the PIM concept is more akin to an employee owned business that gives equal shares. Countries can be ranked according to their level of "userization" in the irrigation sector. While the ranking is rough, the trends are real.

At the upper end of the "userization" graph we find the United States, France, and Japan. Irrigation users have largely replaced the state in managing the irrigation sector although the government retains regulatory functions. At the lower end of the scale, where the state continues to dominate most aspects of irrigation and down to the tertiary or even quaternary levels, we find Morocco, India, Pakistan, etc. The strategies that countries have taken in implementing PIM policies may be characterized according to three basic approaches: (1) the rapid "big-bang" approach of Mexico where water users are strongly pressured to establish an organization to replace the government, (2) the "bottom-up" slow approach of the Philippines with a strong focus on organizing and consensus building, and (3) a hybrid approach which adopts a moderate pace, such as that adopted by Turkey.

Is participation always necessary? Doesn't participation interfere with efficient management in some circumstances? Do we have to allow farmers to come into our board rooms and advise us on how to do our jobs? Aren't there some natural limits to what irrigation professionals should be responsible for and what farmers should become involved in?

A good rule of thumb is that a participatory dimension is important to all management functions. Perhaps there are exceptions to this general rule, but within the field of irrigation management, it is difficult to imagine any. This does not mean that a farmer's council has to be consulted before any decision is taken. If the water availability is so small that only 40% of the demand can be met along a given canal, do farmers need to be asked if they want the water? However, the farmers who receive only 40% of their demand do need to know about overall water availability so they can plan their response, and perhaps suggest better ways of utilizing their reduced share.

"Userism" as a Management Type

We may broadly classify management relationships into three kinds: the first is public management such as the irrigation department. The second is private management such as the Continental Corporation which produces Sparkletts mineral water. A third type of management is neither public nor private in the usual sense. We may call this a users' entity, such as a water user association. To describe this type of entity, we may use the term, "user-ism". You will not find this word in any dictionary; it was coined by Mr. Asif Kazi, Special Secretary in Pakistan's Ministry of Water and Power. We have adopted the term because it captures in one word the process of transferring management from the public sector to organizations of users.

Among these three basic types of management, as applied to irrigation systems, the most rare type is private management. This is mainly because irrigation water is a social good involving large numbers of small farmers, and it is very difficult for a commercial company to manage it with profit. This type of management is clearly not a general option for the irrigation sector. What about management by the public sector? While this is the most common type of management that we see today, in most cases public management has low efficiency and requires substantial subsidies. Experience from many sectors, including irrigation, tells us that it is almost impossible to bring public management into high levels of efficiency. Certainly it is possible to improve the management of public irrigation systems, but it is an uphill battle. The interests of the public managers are unlikely to coincide with the interests of the actual users.

The remaining management option is management by users, or participatory irrigation management (PIM). Under this situation, the managers have a direct incentive to manage the irrigation system efficiently because they are themselves users or are directly accountable to the users. This is the logic of userism: we can ensure a coincidence of interests between managers and users because the users are themselves the managers, or the employers of the managers.

Implementation Strategy

The opportunities for participation are different in each phase of the project cycle. Much of the emphasis on PIM has focused on participation in O&M, and particularly in the recovery of O&M service fees on behalf of the irrigation agency. While this aspect of participation is of great practical importance, there are many ways other aspects of irrigation management where participation can be incorporated. These include: (1) participation in irrigation project identification, planning, and design; (2) participation in system layout and construction; and (3) participation in project monitoring and evaluation.

In short, any aspect of irrigation management can have a participatory dimension. We have discussed why participation us important. In this section we will consider how to achieve it: how to implement participatory irrigation management. There is no recipie for this; indeed, the process of formulating a strategy that fits the specific features of any given country is the first — and ongoing — step. There are some common issues to consider, however, of which we will discuss two:

(1) creating an enabling environment, and

(2) start-up, pilots, and expansion phase.

Creating an Enabling Environment:

- Willingness and interest of stakeholders
- Strengthening interest and commitment
- The role of a PIM "promoter"
- Building consensus about the need for change
- Building Consensus among Farmers
- PIM as one part of a broader "package".

Interest and Willingness of Stakeholders: For participation to work, the government, the incumbent power broker and major "stakeholder" in most national irrigation sectors, must be willing and interested. At least three sections of the government must be willing and interested to support PIM: Political leadership, Administrative leadership, Irrigation agency leadership.

A second, and some would argue more powerful stakeholder must also be brought into any discussions of PIM policies at the earliest stage. We refer here to the farmers themselves, who have the most to lose — and also to gain — from changes in the way their irrigation systems are managed. The membership of farmer stakeholders overlaps with governmental stakeholders in the form of political leaders who represent farming constituencies. These political forces who can speak on behalf of both government and farmers, can be particularly important in both designing and promoting PIM reforms.

Strengthening Interest and Commitment: Expressions of interest in PIM on the part of these entities can be strengthened by facilitating participation in workshops where international experiences are shared (e.g., WBI seminars in Mexico, 1994; Turkey, 1996; Tokyo, 1997; IIMI/FAO seminars in China, 1994; Thailand, 1996). Specific study tours to "model" countries and schemes can be arranged. The World Bank had arranged such study tours to Spain, Mexico, and recently to Argentina and Chile.

Teams of political and administrative leaders have participated in these visits. The first such visit to Mexico by Turkish senior policy-makers was followed by study tours of several contingents of government officials from the Public Works Department. These visits and the shared experiences clearly contributed to the speedy implementation of the PIM program in Turkey. In Egypt, a tour of the USAID-assisted Irrigation Improvement Project was organized for legislators by the Ministry of Water Resources. The tour contributed to a much better understanding of, and subsequent policy support for, shared financing and management of the irrigation system by the government and water users. Methods and tactics for promoting PIM is explored in detail in another section of this Handbook.

The Role of a PIM "Promoter": One clear lesson from participatory experience is the importance of a PIM promoter or "champion" who is effective in mobilizing support within the government/irrigation agency. Typically this role is played by a key official within the agency. In the Philippines, for example, the Assistant Administrator of the NIA is rightfully considered the father of participation in that country's irrigation sector. In Mexico, there were well placed champions within the national water agency (CNA). Perhaps it is possible to rely on a senior consultant (who has the ear of top officials) to mobilize support within the agency. This is the approach being tried in Orissa, India, where a consultant on "farmer organization and turnover" will serve as a guide, but not a direct supervisor, to irrigation department field staff re-trained as organizers. A cadre of specialized social organizers will also be involved as assistants to these engineers. The job profile of the consultant on farmer organization and turnover is given in Annex One.

Building Consensus about the Need for Change: Stimulating a policy dialogue about the need for change and options to be considered can be based both on experience within the country, and by examples from outside the country. Within the country, public discussion of problems faced by farmers whose water services are unreliable or inequitable could be complemented by highlighting other cases where management reforms have resulted in improvements. Universities and centres of higher learning can help organize public discussion events. The media can help in reporting the outcome of these events and highlight key issues in the debate. NGOs can assist in seminars with water users. In Egypt, a professionally-made video film was made available to television stations that attracted large rural audiences in a pilot project area.

Bringing policy makers into contact with PIM cases in other countries is a high priced activity, but can be a powerful ingredient in swaying long-held opinions. Study tours, if carefully arranged and if the right people are involved, can make dramatic differences in the outlooks of individual officials. When Turkey was considering reforms

Building Consensus Among Farmers: A danger in designing and implementing PIM programs (and a cause for their failure) is the lack of attention to farmer interest and support. Since the impetus for PIM oftentimes originates in the perilous state of irrigation agency budgets, the focus for PIM from the agency's perspective could be quite limited, e.g., targeting financial contributions from farmers. The question, of course, is: why should farmers be interested in organizing themselves, if only to pay higher fees? Improvement in services and potential for income enhancement are better motivators for user organization. Farmers' incentives cannot be assumed, but rather must be assessed through field interviews and discussions with a representative sample of the community concerned. Farmers' involvement in designing a management model for PIM which builds-in strong incentives is a critical and often neglected step in the overall PIM implementation process. These issues are explored in another section of this handbook.

PIM as one Part of a Broader: Farmers are interested in much more than just the irrigation system; they want to improve their agricultural production, and more broadly, to improve their livelihoods. In Mexico, when the National Water Commission first met with farmers to discuss their interests, and what their priorities would be for irrigation improvements, the Commission expected that the farmers would ask for more lining of their canals. This is an expensive but important improvement to canal networks in many areas. The Commission was surprised to hear from the farmers that what they were most interested in was full management autonomy. Instead of a physical improvement, they were looking for an institutional improvement. Both types of improvements were within the authority of the government to arrange, but through different means; one required a simple financial outlay; the other required legal changes.

Start-up, Pilots and Expansion Phase

The design of an implementation strategy involves planning for the start-up, piloting and expansion phases of a PIM program:

- *Start-up.* To begin with, management responsibility for PIM must be assigned at the strategic management as well as the operations management levels. At the strategic level, a steering committee could be formed of key officials in the respective ministries, say, public works/irrigation/water resources, agriculture, planning, and finance. This group would approve of the PIM program — its goals, strategies, and a specific work program and budget on an annual basis. At the operations level, responsibility for PIM would be assigned to a specific senior manager. Experiences as to who should be assigned the responsibility vary among countries. In some cases, a senior manager is designated exclusively for Farmers' Organization. In others, a senior manager for Planning or O&M is assigned the additional responsibility. There is no formula for the assignment but whomever is chosen should be interested and committed to PIM ideas. This senior manager would need a group of interested staff from various levels, and representing social science as well as engineering, to help him implement the program.
- The start-up phase would also need resources for preparatory work for the PIM program. In-house expertise may be adequate for this purpose, but often external consultancy inputs may be required. A budget to prepare a pilot three-year program would be a good start and would have to be secured. Sources of financing could be the government's own funds or other external agencies such as bilateral and multilateral donors.
- Piloting. PIM is usually not a new concept to the region or the country. Farmer managed systems exist in most countries. Nevertheless, what is usually new is PIM in all stages of the project cycle, from planning and design to construction and O&M or PIM in the form of irrigation management transfer. In this context, it is useful to talk of "experimenting" with or piloting the PIM idea to test the appropriateness of the various PIM elements to local conditions in the country.

- Yet because PIM implies a new paradigm for the government's relationship with farmers, piloting is not possible in the conventional sense of the term. The irrigation agency must establish a commitment to the basic principles of PIM as a precondition for success in the pilot. The purpose of the pilot should be to help inform the agency how best to operationalize PIM policies, and to demonstrate what kinds of PIM approach would be feasible. It should not be seen as a "wait-and-see" test of whether PIM is a good idea. Inevitable, there will be an internal demonstration function of any PIM pilot, but the management of the agency, at the very least, should support the concepts of PIM at the time the pilot is launched.
- ? The objectives of piloting are to learn from experiences on a small scale so that a manageable irrigation system or subsystem can be the focus of implementation, monitoring, and learning. The implication is that the pilot projects should cover a range of conditions and be carefully monitored so that changes can be introduced in the original PIM model. For instance, in Egypt, 6 pilots covering about 70000 acres were launched in various parts of the country in the late 1980s. These experiences were evaluated in the early 1990s so that the lessons could be used in another set of irrigation schemes serving a command area of about 250000 acres.

Selection of pilots needs careful consideration. Selection criteria would normally include:

- interested farmers in the area
- relative lack of conflict among farmers or a keen desire to end the conflict
- water delivery system that functions relatively well
- tangible improvements can be demonstrated in a relatively short period of time
- supportive irrigation O&M field staff enjoying good rapport with farmers
- local government officials do not oppose the pilot
- size of the pilot scheme is neither too large (and therefore unmanageable) nor is it too small (results have no visible impact on the rest of the government-managed schemes).

To ensure their relevance, pilots must be part of a phased program that includes start-up, pilots, and expansion.

Thus, the Mexican program aimed at the management transfer of all of its large scale irrigation systems (below the secondary canal level) covering nearly 3 million ha in a phased program, learning from experience during implementation.

All of the phases should ideally be designed before pilots are implemented, thus placing them in the overall context of a country's institutional reform program. An additional problem with pilots is that they attract a great deal of resources, sometimes disproportionate to the size and scope of the experiment. Replicability becomes a casualty.

Expansion: At the end of the pilot phase, policy makers and managers would have derived lessons regarding the scope for user participation in the management of water, funds, operation and maintenance and capital development of the system. They would also be wiser about current government policies and procedures that help or hinder effective participation.

The period before the large-scale expansion phases presents an opportunity to bring about changes in existing laws, policies, financing arrangements and procedures as the need may be. Without this intervention, an expanded program could run into severe problems. What was possible in an informal, small scale setting during the pilot phase may not be easy during the expanded large scale phase.

In addition to required changes in current policies, agency managers have to face the question of institutional capacity — trained staff, budget, managerial resources — to undertake a large scale expanded program that would cover a larger geographical area. Some time may have to be initially spent during the expansion phase in strengthening institutional capacity to manage the effort. These issues will be further explored in the section on Training Strategies.

Training Strategy

What Training do the Organizers Need?

Assuming that agency staff will be used for the organizing, and assuming that their professional training has been in irrigation engineering, they would need to be re-tooled as organizers. First of all they would need a thorough understanding of the rationale for PIM, so that they can present a clear message both to the farmers and to their own colleagues within the irrigation agency. Secondly, they will need training in communication skills (including listening skills) for effective interaction with the users. Thirdly, they will need training in social analysis, including an understanding of social stratification (by caste, ethnicity, or class), kinship, patron-client relations, labour relations, religious factors, political affiliation, land tenure (tenants, share-croppers, owners), etc. And fourthly, they would need training in methods for gathering information from farmers (e.g., participatory rural appraisal) and in methods for organizing farmers.

Legal Framework

The legal framework for the establishment of WUAs, and for enabling them to operate and maintain such parts of the irrigation system, consists basically of three sets of legal instruments, namely:

- The enabling law,
- The bylaws of the WUA, and
- The transfer agreement between the irrigation agency and the WUA.

The existence of a basic law on water, or on WUAs, is certainly an important parameter for the other parts of the legal framework because the basic law would usually specify the main issues that need to be included in the bylaws and transfer agreements, and would also determine the manner in which those issues are addressed. Those issues would usually include the procedure for establishing WUAs, the rights and duties of the WUA and the irrigation agency and the relationship between them, and the structure of the water rates and other fees.

The Enabling Law

For a WUA to be established as a legal entity, there has to be a law authorizing its establishment. This law could be a general comprehensive "Water Law" that deals with all aspects related to water, including establishment of WUAs. The National Water Act in Mexico, and the Water Resources Act in Nepal, are examples of such a comprehensive law. The enabling law could also be special rules and regulations dealing specifically with WUAs, and deriving their authority from a basic law, such as the "Implementing Rules and Regulations on the Provisions of Republic Act No. 7607" on small farmers in the Philippines.

Because of the absence of a basic law specifically on water or on WUAs, the states of India have relied on different laws to establish WUAs. In the Indian State of Maharashtra, WUAs have been established and registered as co-operative societies under the "Co-operative Societies Act." On the other hand, in the State of Tamil Nadu and the State of Orissa in India, WUAs are established and registered as societies, under the "Societies Registration Act."

The law establishing WUAs would usually include provisions indicating that the WUA to be established is a legal entity. Such enabling law would also address the relationship between the WUAs and the irrigation agency, the duties and obligations of the irrigation agency, and those of the WUAs, and the structure of water rates and the operation and maintenance and other fees. The enabling law may also lay down some of the main issues to be addressed in the bylaws of the WUA, and in the transfer agreement.

All societies, including WUAs, registered in any state in India under either the Co-operative Societies Act, or the Societies Registration Act, are legal entities capable of contracting, opening and operating bank accounts, and instituting and answering suits. However, because of their general nature, those two Acts do not address certain WUA related issues that are usually addressed in special water law or through a law establishing WUAs. It is for this reason that specific legislation dealing specifically with WUAs is often desirable.

Bylaws of the Water User Association

Whether established under a separate law or under an umbrella enabling law, the WUA would normally be required to prepare and agree on its bylaws before it can be registered as a legal entity. Those bylaws may be called "Regulations," "Constitution," "Charter" or "Articles of Associations." The issues that such bylaws need to address include:

Basic Facts About, and Objectives of, the WUA: The basic facts would include the name of the WUA, the law under which it is registered and its registration number, its address, and a clear definition of the area that the WUA is serving or its area of operation. This area of operation could be an entire irrigation district, or an entire command of a distributary, minor, sub-minor or a water course. It could also be defined by its size in acres or hectares. A broad statement on the objectives of the WUA is usually included in the bylaws. Such objectives would include: participation in the management, operation, maintenance and upgrading of the irrigation infrastructure works that the WUA has taken responsibility for, collection of water charges, and provision of irrigation and drainage services to the members of the WUA.

Criteria for Becoming a Member of the WUA: Most bylaws restrict membership of the WUA to the registered land owners in the hydraulic unit, who are engaged on a full-time basis in farming. If any member of the WUA sells his land, his membership will be automatically cancelled, and the new owner will be eligible for the membership of the WUA. However, the bylaws in some countries extend the right to become a member to both owners and tenants. In the Indian State of Maharashtra membership of the WUA is extended to "Any owner/cultivator/permanent tenant/protected tenant in the area of operation of the society...." These bylaws extend membership of the WUA beyond owners and tenants to include other categories of users such as sharecroppers and encroachers who are prevalent in some parts of India.

Number of Farmers Required for the Establishment of a WUA: Most of the bylaws state that at least 51% of the registered land owners in the command area where the WUA would be established should be enrolled as members before the WUA can seek registration, and before it can be allowed to operate. However, some bylaws allow the WUA to seek registration based either on the number of farmers enrolled, or the size of the land holdings coming under its operation regardless of the number of farmers enrolled.

In Orissa "At least 51% of the registered land owners in the command area covered by the Association should be enrolled as members, or the land holdings of the members should cover at least 51% of the total area under the proposed Association." Moreover, in some cases, once 51% of the farmers are enrolled and registered in the WUA, all other farmers within the command area will be deemed to have become members in that WUA. In Mexico "all users listed in the register who, while not founder members of the Association, apply and pay for irrigation services, thereby tacitly agreeing to belong to the Association, shall also be deemed to be members with the same rights and obligations."

The WUA as a Legal Entity: Although the enabling law would usually specify that the WUA is a legal entity, further details regarding what this entails are usually included in the bylaws. It is usually stated that the WUA is authorized to enter into contracts in its name, and that the WUA can sue in its own name, and answer suits instituted against it. The WUA can also be authorized to borrow funds from private sources, using, if necessary, its assets as a collateral.

Structural Organization and Internal Management: Although WUAs may be organized differently, the two most common ways of organizing a WUA at the membership level are:

(i) a general body of the WUA which would consist of all registered members who are current in the payment of their dues, as in Mexico, Nepal, and the Indian states of Maharashtra, Orissa and Tamil Nadu; or

(ii) a general body, which could be called the general assembly, which would consist of delegates directly elected to represent the different irrigation districts or sub-units within the hydraulic unit, as in the case of some WUAs in Turkey. In the latter case, there is an absence of one single forum encompassing all members.

The executive body would usually consist of a president (or chairman), a vice president (or vice chairman), a secretary, a treasurer, and other specified number of members, and those posts do not, in many countries, carry remuneration. One exception is Turkey where the president of the WUA is paid a salary, and the other members of the executive body are paid honoraria.

Membership of the executive body may also be extended to non-governmental organizations (NGOs) or any other institutions interested in irrigated agriculture, including a representative of the irrigation agency, but such members usually do not have the right to vote.

A bank account in the name of the WUA is usually opened, with separate sub-accounts for water charges, operation and maintenance fund and membership fees. The executive body would also approve work expenses, engage labour, organize labour contributions from members, keep systematic accounts and records of amounts collected and those spent on the water charges, operations and maintenance fund, and membership fees. The executive body is usually required to have such accounts and records audited annually, and submit such accounts and audit reports to the general body during the annual meeting for approval. The executive body could include members who are specifically designated as auditors.

Operation and Maintenance: One of the primary objectives of the WUA is to operate and maintain the transferred irrigation and drainage system efficiently and economically, and with the full and active participation of all the members. Operation would include receiving water in bulk from the irrigation agency at a prescribed rate at the head of the minor/distributary and distributing such water equitably and in a timely manner, as per procedures and criteria agreed with the irrigation agency, to all farmers in the hydraulic unit, whether members or non-members. The bylaws would lay down, in agreement with the irrigation agency, the criteria for allocation of water to both members and non-members, which could be based on the type of crop grown or the size of the area to be irrigated, or both. The bylaws would also include the criteria for assessing water charges and operation and maintenance charges from both members and non-members. The operation and maintenance fund could include sources other than charges from farmers.

Water Charges: Reference to water charges is usually included in the enabling law.

Other details on water charges are also included in both the bylaws and the transfer agreements. Inclusion of provisions on water charges in the bylaws would serve the purpose of establishing the payment obligations of each member of the WUA, whereas the provisions in the transfer agreement would establish the payment obligations of the WUA, as a legal entity, vis-a-vis the irrigation agency.

The transfer agreement would usually include provisions on the manner in which water charges are calculated, and the due date or dates for payment of the water charges by the WUA to the irrigation agency. It may include provisions for the payment of commission or discount by the irrigation agency to the WUA on water charges collected by the WUA. The transfer agreement would also include provisions giving the irrigation agency the right to suspend delivery of water if the WUA fails to make the payments of the water charges within the prescribed or extended time limit. Moreover, non-members of the WUA may be required to pay higher water rates than those paid by members.

Rights and Obligations of Members: Under most WUAs, each member of the WUA has one vote regardless of the size of his land holding. The bylaws would also need to address the issue of proxy voting-whether it is allowed, and if so, the maximum number of proxy votes one member may cast on behalf of other members. Failure of a member of the WUA to meet his membership obligations as described in the bylaws, such as failure to make payments, permit inspection of the irrigation system in his land, comply with the terms of the transfer agreement, carry out proper maintenance, or allow delivery of water to other users may subject such a member to sanctions. Such sanctions may include suspension of such a member until all outstanding obligations are met.

Interpreting and Amending the Bylaws: Provisions would usually be included in the bylaws themselves describing the procedures for interpreting provisions of the bylaws in case there are different views as to what a certain provision may mean. Procedures and quorum required for amending the bylaws would also be included in the bylaws. Usually interpretation of provisions of the bylaws would be referred to a central body such as the Registrar of Societies, or the irrigation agency, and amendments would be effective after approval of such body.

Establishment of a Federation of WUAs: The bylaws of some of the WUAs, as in Mexico, Colombia, Tamil Nadu and Orissa, include provisions enabling the establishment of a federation encompassing registered WUAs in one command area. Those bylaws specify the responsibilities of the federation, and the relationship between such a federation and each of the affiliated or member WUAs. Usually, the presidents of each of the WUAs that decided to establish or join such a federation would represent their WUA in that federation. The federation would have an advisory non-binding role over the member WUAs, and may be used to resolve any disputes among such member WUAs, or between WUAs and the irrigation agency.

Bylaws of the Water User Association

The transfer agreement is the agreement between the WUA and the irrigation agency in which the irrigation agency agrees to transfer to the WUA responsibilities for operation and maintenance of certain parts of the irrigation system, including the drainage system, and the collection and remitting of water charges; and the WUA agrees to carry out such responsibilities. This agreement may also be called "Memorandum of Understanding (MOU)," "Transfer Protocol," "Concession Agreement" or just "Concession." The issues that such transfer agreement would need to address include:

Area and Irrigation System to be Transferred: The agreement would need to define clearly the irrigated area to be transferred, specifying the size of the area, and the command under which it falls, and including the irrigation system existing there that is being transferred. The system to be transferred is usually the irrigation system at the level of primary, secondary and tertiary, including the drainage of such areas too. A copy of a map showing such area may be attached to the agreement. The agreement would specify whether the ownership of the irrigation system, including the land and structures and works there on, remains with the irrigation agency or is being transferred to the WUA, together with the operation and maintenance of such irrigation system. Provisions should also be included clarifying whether the ownership of any ancillary equipment is being transferred.

Interim Joint Management: Some agreements may provide for a joint management of the irrigation system for a short period of time by both the irrigation agency and the WUA. The rationale for such joint management is to prepare the WUA, during this interim period, for taking over full responsibility for operation and maintenance of such irrigation system. During this interim period which may run for up to one year, officers of the irrigation agency would train WUA representatives in compiling necessary data, and preparing and testing operation and maintenance plans for the distributaries, minors and subminors to be transferred to them. They would hold joint inspections to identify any problems in the irrigation and drainage system, and to agree on how to deal with them.

Transfer of the Irrigation System: Transfer of the irrigation system to the WUA will normally be preceded by a number of actions, including the preparation of an inventory of the works, structures and equipment to be transferred, joint inspections of those works, structures and equipment by both the irrigation agency and the WUA representatives, carrying out of necessary testing, and repairs, if any, at the irrigation agency cost, and handing over management of the system, along with all necessary documents and instructions, to the WUA. The WUA would need to satisfy itself that the system is, indeed, in a good working condition, as the transfer agreement would include provisions that the irrigation system, at the time it is transferred, was in good working condition.

Responsibilities of the Irrigation Agency: The agreement would spell out clearly the responsibilities of both the irrigation agency and the WUA. Responsibilities of the irrigation agency would include handing over the system in a reasonably operating manner, delivery of water to the WUA in bulk at the agreed time and providing the WUA with any agreed upon financial assistance and other benefits. The agreement could include provisions absolving the irrigation agency from liability should it be unable to deliver the agreed upon amount of water, or unable to deliver it at the agreed time, for reasons of force major, or act of God. In periods of water scarcity or emergency, after the demand for domestic and other priority uses is satisfied, the irrigation agency would usually have the authority to decide that the remaining water for irrigation shall be allocated to crops of utmost importance to the community there.

Responsibilities of the WUA: A number of the responsibilities of the WUA detailed in this section of the agreement are usually spelled out in the bylaws of the WUA, but may still be included in the transfer agreement to clarify the obligations of the WUA towards the irrigation agency. Such responsibilities would include: operating and maintaining the irrigation system transferred to it, including the drainage system, in a proper and satisfactory manner; receiving water in volumetric basis, and distributing such water equitably and in a timely manner, based on clearly defined criteria, to both members and non-member farmers in the operation area, and collecting the water charges agreed with the irrigation agency.

The responsibilities also include establishing the operation and maintenance fund, and maintaining and repairing, in a satisfactory manner, all the field channels, field drains, minors, subminors and distributaries, together with the structures there on in the operation area of the WUA. In addition, the WUA may be responsible for the maintenance and repairs of any equipment and machinery transferred to it. Such equipment and machinery may be transferred to the WUA as part of the irrigation system for which it is now responsible, or may be separately leased by the irrigation agency to the WUA at an extra cost.

The WUA would also be responsible for the security of the infrastructure transferred to it, and such responsibility could either be carried out by the members of the WUA themselves, or through hired labour. Maintenance would usually include silt clearance and removal of weeds from all water courses under the WUA. It would also include earthwork to restore banks and repairs to other structures, in addition to maintenance of service roads. Usually minor repairs are carried out by the WUA and major repairs by the irrigation agency. Definitions of what is "minor" and what is "major" should be included in the transfer agreement (repairs to damage caused by natural disasters such as heavy rains, floods or earthquake are usually considered major repairs). The WUA may be required to prepare an annual maintenance program for the irrigation system under its responsibility, including any machinery and equipment, and to submit such program to the irrigation agency for approval prior to implementation. Operation and maintenance of the irrigation and drainage system, other than the one transferred to the WUA, would continue to be the responsibility of the irrigation agency.

The agreement would authorize the irrigation agency to suspend supply of water to the WUA if maintenance and repairs were not being carried out properly, or to carry out the repairs itself and recover the cost from the WUA. The agreement would also include provisions on how disputes between the WUA and the irrigation agency, arising in the course of the operation and maintenance of the transferred irrigation system, would be settled. Such disputes could be referred to a committee comprising one representative from the irrigation agency and the water users' federation.

Termination of the Transfer Agreement: Although the enabling law may include provisions on the termination of the transfer agreement, usually more detailed provisions are included in the transfer agreement itself. The agreement terminates after expiry of the number of years specified in the transfer agreement, which may be as high as twenty years as in Mexico. However, the agreement would usually be subject to renewal for another similar period. Moreover, failure by the WUA to comply with the provisions of the agreement, including the failure to properly operate and maintain the irrigation system transferred to it, or to make timely payment of water charges, or to take corrective measures within a specified period of time, as agreed with the irrigation agency, would give the irrigation agency the right to terminate the transfer agreement.

Organizing Processes

PIM implies the establishment of an organization of water users. In cases of management transfer, the establishment process focuses on a level of user organization where no organization presently exists, that is, at a level of the irrigation network previously managed by the government irrigation agency. Typically this entails organizing federations of user groups at the tertiary level under a secondary-canal association. The challenge of creating a new organization of users is perhaps the most central feature of the management transfer process. The act of management transfer from the agency to the users depends upon a user organization that is capable of assuming those management responsibilities.

Before any organizing of the users is carried out, there needs to be a package of incentives in place for both the users and the agency staff whose jobs would be affected by the transfer program. Such incentives are needed both to make the program work, and to maintain credibility with farmers whose long-term support will be required. If organizing is attempted before an adequate incentive structure is established, the transfer program could well collapse, thus setting the entire program back by several years, as well as causing short-term hardships to those concerned. Thus, if the incentives are not clear and attractive to farmers, the organizing process should be delayed until the incentives are clarified.

First determine:

1. Who will do the organizing?
2. What training do the organizers need?
3. Who will provide the training?

Then, the organizing steps for transfer of management to users.

Who will do the Organizing?

The first step is to decide on the type of organizers who will work directly with farmers in helping establish the organization. [We use the term "water user association" (WUA) to refer to this new organization, although the term, "WUA" can also refer to organizations at the tertiary level of the system which have always been outside the management control of the government irrigation agency.] In the Philippines, a special cadre of social organizers was recruited and trained by the National Irrigation Administration in the late 1970s and early 80s. These organizers were mostly social workers or social scientists selected for their ability to work easily with farmers in village conditions. They were trained in irrigation management so they could better understand the technical problems of the farmers they were trying to organize.

The work of the Philippines social organizers was effective, yet this approach has not been widely replicated by irrigation agencies in other countries. The investment and bureaucratic difficulties involved in recruiting new temporary staff from a different discipline has led many irrigation agencies to try other approaches. In Mexico's transfer program, the National Water Commission's own staff were used along with staff from a sister agency, the Institute for Water Technology (IMTA). In addition, a few consultants were brought in on a case by case basis. In India, some state irrigation departments have relied on extensionists from the Agriculture Department under the Command Area Development Program. Also in India, several NGOs have been involved in organizing, at the request of the irrigation departments. But the organizers of choice, for most irrigation agencies, will be their own field staff. These staff are already within the bureaucratic structure of the agency, so the lines of authority are clear, there is little additional expense involved, and these staff are already familiar with the physical systems and with the local farmers.

The problems with using irrigation agency field staff for organizing work, however, are considerable. These staff are not necessarily interested in organizing farmers, They have not received any prior training in social work (in most cases), and their superiors have also no training (nor interest) in these tasks. These would-be organizers must be re-trained for their organizing tasks, and just as importantly, their job assignments need to be redefined to reflect their new role. In addition, their superiors need to be trained and re-oriented so they understand and appreciate the new role to be played by their field staff.

Organizing Steps

Note: This assumes there are reasonable incentives for farmers to take over system, and for the government irrigation agency to hand over the system; if not, go back and work on the incentives!

- Organize the organizers: Arrange for supervision/support for the organizers, and clear lines of communication with Departmental staff responsible for the overall project (Ensure that PIM component of program is well integrated with rest of project).
- Meet the farmers and other irrigation stakeholders/Discuss plans formulated during participatory design phase:
  - — village head
  - — local administrative officials
  - — local political leaders (MLAs/MPs)
  - — leaders of other farmer organizations (producer cooperatives).
- Identify key power relations among farmers; develop strategy for organizing
- Establish provisional boundaries of the system (through consultation with key power brokers among the farmers); conduct inventory of potential members; draw map showing command area and irrigation system.
- Arrange series of meetings between farmers and Departmental field staff to discuss improvements that need to be made prior to handover.

  - arrange canal walk-through to discuss specifics of design/ infrastructure improvements
  - discuss general terms of WUA contracts.
- Arrange farmer visits to other associations to discuss with those farmers (and invite those farmers to visit new association).
- Organizational Assistance:
  - Help prospective WUA leaders arrange farmer meetings to discuss plans
  - Help formulate/revise bylaws
  - Advise on elections/selections
  - Assist with legal registration
  - Arrange management training for WUA leaders.
- Arrange meetings between WUA leaders and Department staff to discuss details of WUA contract and terms of transition phase leading to hand-over;
- Advise on staff recruitment(?)
- Assist with formal hand-over
- Visit periodically to monitor WUA's performance.

Reorienting Agencies

Agencies implementing PIM need to reorient themselves to a new style of water and infrastructure management. No longer is the agency the sole manager of the water system; rather, the agency staff become management partners with farmers. The agency retains control over the water supply and head works (in most cases) and perhaps the main canal network. Their management must be closely coordinated with the management of the lower levels of the system, now under the control of user associations. Under this new arrangement, the agency has both a direct management role (for the highest levels of the system) and an indirect management role — facilitating and supporting the work of the user associations.

- Restructuring the Irrigation Agency
- How agencies do NOT Change under PIM
- Facilitating and Hindering Factors.

Restructuring the Irrigation Agency

Even without an explicit PIM program, changes over the past two decades in the irrigation and water policy context clearly point to needed changes in agency roles and functions. Irrigation agencies are being forced — by citizen concern as well as by the conditions imposed by international financial organizations — to pay a great deal of attention to social and environmental impacts of irrigation activities. These have become important aspects of project assessments in the 1980s. Similarly, as user participation has gained importance, irrigation agencies have had to re-evaluate their roles and functions with respect to irrigation development and management.

Incorporating PIM requires a range of new functions and new organizational structures for the irrigation agency; it must shrink and shift to make room for user management. It must shrink in staff and functions, and it must shift from directly doing O&M to indirectly supporting O&M within the user managed level of the system. Specifically, changes in agency roles require the agency to perform new tasks as explained below.

- Revised Structure. At least some field-level staff of the agency, and perhaps most field staff, would be replaced by the new staff of the WUAs. As a general rule, higher level staff of the agency would be less affected by the transfer of O&M responsibilities.
- Consultation. Staff in the various functional departments of the agency need to commit to and gain expertise in farmer consultation.
- for those staff in planning and design, PIM implies a consultative step that would provide farmer inputs into prioritising system improvements, selecting designs and layouts, and determining outlet locations. On the part of agency staff and their consultants, consultation with farmers for planning and design requires listening skills and an ability to present technical details in simple language. It also implies the flexibility to change plans and designs based on farmers' suggestions.
- for O&M staff, PIM implies the ability to respond to feedback on system performance from water users. Field staff would need to reorient themselves from carrying on as operating engineers to performing advisory functions — a role in which they may have little experience or expertise.
- Accountability. The accountability of irrigation staff increases considerably with PIM and the formation of WUAs. Both the agency and the WUA need to maintain accurate records of

water delivery (planned and actual) to ensure that the agreed amount of water is delivered/received. Where PIM is introduced at the survey and design stage itself, information on initial designs and comparative costs would have to be shared with water users. The importance of shared information increases significantly with PIM, since now agency staff are not only providing information "upwards" to their bosses but also "downward' to their clients, the managers of the WUAs.

- Advisory Services. Agency staff need to reorient themselves to extend technical and managerial advice to water users. PIM is effective only when informed water users participate knowledgeably in the planning, design, construction and management of irrigation systems. In Egypt, where water users are organizing themselves around a common pump at the head of the tertiary canal, the Ministry of Public Works and Water Resources has established an Irrigation Advisory Service to extend assistance to WUAs.

How Agencies do not Change Under PIM

Many of the irrigation agency's core functions continue after PIM is introduced. In most cases of PIM, the public irrigation agency maintains the legal responsibility for overseeing the wise use of the irrigation facilities. Under most contracts, the WUA is managing these facilities for a set period of time (25 year concessions in Mexico). Cases of outright privatisation — where the government agency permanently sells or cedes the ownership of the infrastructure to a WUA — are very much the exception. Thus, the public irrigation agency has not only a right but an active responsibility to regulate, for example, groundwater levels and soil and water quality.

Management functions also continue, with the agency maintaining responsibility for the head works and (usually) main system. The agency also has a monitoring role in ensuring that the WUA is adhering to the contractual agreement of management transfer, and is properly maintaining the infrastructure and adequately delivering water to the WUA members.

Facilitating and Hindering Factors

Incentives within the agency are often opposed to the establishment of strong WUAs. Organized users pose a threat to the agency's control and to the endemic problem (in many countries) of bribes and kick-backs on contracts. Management transfer programs which aim to expand the role of WUAs threaten the jobs of agency staff. And the task of mobilizing the formation of WUAs becomes an added responsibility for these same staff, and not a task in which they have either professional training or interest. Finally, because irrigation agencies are engineering organizations, the professional rewards and recognition are in design or construction of physical schemes, not in dealing with farmers' demands.

Given this situation, which naturally varies considerably from country to country, how could PIM possibly gain a foothold? How can an agency transform its own management culture? There are a number of tools for doing this:

- Performance evaluation criteria: Including work with WUAs in the job descriptions and performance evaluations of agency staff is one way to create incentives for PIM. Supervisors (whose own job descriptions also need to incorporate PIM criteria) need to monitor their subordinates' performance in working with WUAs, and request feedback from farmers on how helpful staff members are. Publicizing the performance plans of agencies among users may strengthen agency and staff accountability. While the number of associations registered is relatively easy to measure, it is not an adequate indicator of staff performance with WUAs because it does not take into consideration how well the WUAs and joint management work. Fee collection rates from farmers or from WUAs may provide a better indicator of farmer satisfaction.
- Monetary and non-monetary rewards: Monetary rewards are often difficult in the context of civil service regulations. In some cases, where the agency is an autonomous corporation, a reward system may be possible. The National Irrigation Administration, a semi-autonomous agency in the Philippines, was able to offer higher than regular civil service salaries to its staff. However, this was contingent upon maintaining a balance between expenses and cost recovery, which gave staff an incentive to work with WUAs to reduce agency costs and raise irrigation service fees. Non-monetary rewards can also provide a powerful motivator for agency staff. In Nepal, a special medal was awarded to an Irrigation Department engineer who had made extraordinary efforts to establish a WUA. Some already motivated staff will derive job satisfaction from challenging new tasks such as WUA organization, provision of technical assistance, monitoring and regulation,

rather than routine O&M. But the agency's management must provide clear and consistent signals that such new roles are valued.

- Reduced transaction costs. Reduction in the number of conflict cases (or "hassle factors") to be resolved by the agency staff has also proved to be a valuable incentive for working with WUAs. The establishment of WUAs as an organized forum for communication can reduce the transactions costs for agencies and for farmers. Reduction in damage to structures, as farmers focus on protection of system facilities eases the burden placed on agency field staff. Thus staff may have to devote less time to field inspections and user management often results in greater vigilance over infrastructure and fewer complaints. Thus, there may be less need to deal with individual farmer demands as WUAs take on additional roles. In Chile, problems with the state management of irrigation are no longer "politicized" by farmers since users' associations and organized forums for articulation of problems have been established.
- Skills Training. Preparation is needed for two types of activities that could result from the introduction of PIM:
- Activities dealing with user participation, including farmer organization, WUA establishment, technical assistance for WUAs, fostering agreements with informal and formal WUAs, public communication and information etc.
- Activities that can be taken up by agency staff as a result of water users assuming a larger role in irrigation O&M thus freeing up time and resources for the agency.

Agency management will have to formulate a program of short and long term training to prepare the staff for changes in roles and functions and upgradation of expertise. It may not be possible to meet all needs through training of available staff. New staff may have to be inducted into the agency. Secondment from other departments of government, from universities or NGOs may also be considered. For instance, in Egypt, the Ministry of Public Works obtains the services of agricultural extension staff through secondment from the Ministry of Agriculture. In Philippines and in India, community organizers are hired as consultants and contract staff.

Financial Aspects

O&M can be financed directly (through a direct tax based on O&M costs) or indirectly through land taxes or other kinds of agricultural taxes. The relative share of government vs. farmer contribution varies dramatically across countries, with some countries using O&M as a vehicle for delivering subsidies to irrigated farmers.

- How to set O&M cost
- User participation in defining service levels
- Assessing users' willingness and capacity to pay
- Determine appropriate charging mechanisms
- Linking fees and services.

How to Set O&M Cost

The managing agency needs to identify the various services it provides through the water delivery and drainage systems under its management control and then allocate costs to each service category. This information should be made freely available to farmers, and to their WUA leaders. Such information helps WUAs understand the respective contribution of government and farmers and helps the WUA determine which functions it might take over from government, and which functions it might prefer to pay government to provide. An aspect of the cost determination activity is the separation of "staff or administrative" costs and "works or program" costs. A clear rationale for the two types of costs is important in justifying the irrigation service fees which farmers are expected to pay. This is equally true for the water fee payable to the government agency, and to the fee payable to the WUA.

User Participation in Defining Service Levels

Involvement of farmers/WUAs early in project design to determine the level of service required by farmers is crucial. Users' involvement in setting service levels will ensure the genuine demand for those services, which will be expressed by a willingness to pay the fee required to support that level of service. Some examples of users working with the agency in determining what services the government should provide are given in the following:

- In Mexico, following the transfer of management of irrigation districts to WUAs, irrigation plans are prepared by WUA-hired managers at the level of the module, each covering about 5-8000 ha, taking into account cropping plans, conveyance losses, and equitable distribution. These are then negotiated with the national irrigation agency to determine the final allocation, generally based on an arranged demand pattern. O&M costs at the level of the secondary and below are met by user fees managed by the WUOs. In addition, WUOs contribute to a part of O&M costs of higher levels of the system.

- In the Office du Niger, Mali, joint committees have been established for O&M in every region. The committees have 5 to 10 representatives of producers and 5 to 10 representatives of the Office. These committees decide on types of services, costs including procurement matters, and water service fees. They also make decisions on the use of 50 per cent of the user fees collected for O&M.
- In Chile, the national federation of WUOs was consulted in the design of a Bank supported irrigation project. The federation and local WUOs played an active role in project preparation especially in discussions of service options and costs. Subsequently, the project incorporated the condition of WUO approval for investment proposals and other project components.

Assessing Users' Willingness and Capacity to Pay

The readiness of the user to pay will depend not only on the benefits to be derived, but on the level of payment to which the users have become accustom. It is in the farmers' interest to pay as little as possible. But it is also in the farmers interest to receive reliable supplies of water, and for this he will certainly be willing to pay something. In a rehabilitation project in Egypt which is introducing PIM, it has been estimated that users will derive incremental income from the project due to expansion of irrigated area and yield increases and that incremental costs of their participation in irrigation management would amount to about 20 per cent of this new income. Increases in income would also depend on agricultural extension services and these would have to be strengthened to allow farmers derive full benefits from the improved water services delivered by the project.

Financial viability of the WUA enterprise is critical to the sustainability of the organizations and the irrigation infrastructure. It is important to examine the total costs the WUAs have to bear, including staff, materials, travel to meet with government officials, and formal and informal payments that must be paid to government agencies. If the level of fees members must pay to meet these costs (which are often in addition to continued payments to the government) is too high a portion of gross or net income from irrigation, the WUAs are not likely to succeed. This is particularly problematic for pump irrigation systems, in which high energy and maintenance costs exceed what the organizations are able to collect from farmers.

In some cases, WUAs must depend on income from other sources to subsidize their irrigation activities. For instance, some irrigation districts in the Western United States rely on revenue from power sales to balance their irrigation budgets.

Determine Appropriate Charging Mechanisms

There are a number of alternative charging mechanisms available to agencies and WUOs in structuring O&M fees. As a principle, the fee structure should be equitable, administratively simple, and easily understood by both users of a particular service and the staff that will administer and collect the fees.

User Fees: Generally, user fees are charged by area irrigated. The advantage of this method is that it is relatively simple to understand. However, it does not reflect the quantity of water used nor does it provide an incentive to conserve water. Volumetric charges address those concerns. However, there are many measurement difficulties in operationalizing this method in the field. Agencies often favour a combination of the two methods. In Mexico, for instance, the agency charges the WUAs volumetrically at the turnout of the secondary canal. In turn, the WUAs base their water charges to individual members on area irrigated and type of crop. In addition to unit area and volume, WUAs have also resorted to other bases such as charges for the entire season which clearly favour the high volume user. The design of irrigation fees should be a topic in WUA training programs so farmers understand the implications of various modes of charging user fees for efficiency and equity.

Property Tax: This is a charge based on the value of the land receiving the service. Irrigation and drainage enhance land value. The higher the land valuation, greater the service payments. The problem with this method is that generally land taxes are collected by the Treasury and are not transparently linked with irrigation services. Also, the valuation and revaluation process demands a great deal of administrative resources.

In-kind Contributions: Contributions of labour, materials or both by the benefiting farmers towards O&M reduce the cost of services. In Viet Nam, the provincial and national governments finance schemes down to 150 ha for new irrigation development. Below this level, farmers must build the channels, with the government providing support for survey and design, as well as materials in some instances. After schemes are completed and taken over by farmers, each member must provide up to 20 person days per year towards maintenance of the tertiary as well as secondary systems.

Replacement of Assets: A significant cost in irrigation services is depreciation of the asset base used to store, transfer and deliver water services to users. Facilities employed range from minor tools and equipment, buildings and housing, motor vehicles and heavy equipment, canals and drains, pumping stations, and where applicable, major structures including dams. Unfortunately, depreciation costs are usually neglected in estimates of cost of water services whether provided by irrigation agencies or by WUAs. Assets are consumed at different rates and each year, a significant non-cash value is "lost" as assets depreciate, even with adequate maintenance. Without proper maintenance, service life will be shortened and depreciation will accelerate. The agreement between the irrigation agency and the WUA includes O&M standards to be maintained. Provision is also made for collecting and retaining disaster funds by WUAs. In Mexico, the Hydraulics Committee at the district level approves the annual O&M program proposed by the WUA.

Care of assets is vital to the sustainability of water system performance. Maintenance and the ultimate replacement programs for all assets need to be planned and implemented in an effective and economical manner. In parts of Vietnam, the estimate of costs of services (and hence of user fees) include provision for depreciation of assets. Adequate data about the location, age, condition and serviceability of all assets is therefore essential to this process. WUAs need to be trained and assisted in preparation of an asset inventory and in the formulation of a maintenance program.

Linking Fees and Services

An obstacle to user contributions is the perception that their payments are not linked to the irrigation agency or its services. Often, money for services is collected by the central revenue department, and from there funds are allocated in the various government agencies, including irrigation, with little link between revenue earned and costs of service provision. To reverse this process, the money cycle must be transparent.

Ideally, WUAs would collect O&M fees and manage the system under their control and pay a service fee to the agency for that portion of the system under the agency's control. Under this arrangement, the agency and the WUA agree on mutual rights and responsibilities and the WUA is completely responsible for management of water, infrastructure, and finances for a part of the irrigation infrastructure.

If users are to pay service fees with the government managing the system, fee collection is preferably done through the WUAs. The Philippines experience has shown that fee collection is far better in systems where WUAs are organized and where they have a role and incentive for collection. For instance, in the pilot projects in Maharashtra, India, WUOs are allowed to retain a proportion of collections as a bonus. The cost of collection by WUAs is lower relative to government agents. Sanctions for non-payment by individuals are easier when enforced by WUAs. The danger is that considerable time and energy are devoted to fee collection on the part of the agency staff as well of the WUAs and fees rather than reliable water services becomes the focus of management attention.

Where users pay service fees to the managing agency, fees are a good means of signalling satisfaction with services, as in the case of Indonesia. Such information is publicized and used for evaluation of agency services and management. In any case, transparent accounts and audits are fundamental in showing WUA members that the financial management of the organization is sound. General body meetings of the WUAs must discuss financial performance including the accounts and audits.

WBI Training Programme

In most developing countries, irrigation development projects and their operation and management are heavily dominated by the public sector. Conventional wisdom has assumed that only the state was capable of handling large modern projects requiring heavy capital investment, complicated technical inputs, the legal mandate to distribute water, and collect fees.

Recent experience challenges these assumptions. Government-operated irrigation systems are often poorly maintained with steadily deteriorating infrastructure. Yet some of these same systems show dramatic improvement when their management is transferred to water user associations (WUAs), which enter into contracts with the government for operating and maintaining portions of the system or in some cases entire systems. Since the mid-1980s, countries including Mexico, Turkey, Indonesia, Philippines, Colombia, India, Srilanka, and Nepal have adopted policies to encourage greater management participation by water users. One of the most dramatic management transfer programs has been in Mexico, where the government adopted a policy to gradually transfer all its large scale irrigation districts to 78 WUAs. As of mid-1993, the management of more than 1.2 million hectares (ha) of irrigated land has been transferred to WUAs.

In response to interest in participatory irrigation management expressed both within the World Bank and in a number of borrowing countries, the World Bank Institute of the World Bank has initiated a five-phase program on participatory irrigation management (PIM). The overall purpose of WBI's program on PIM is to stimulate high-level policy dialogue on participatory irrigation management within the country, leading to policy commitment and programmatic action.

The PIM training program has thus far been initiated in India, Pakistan, and Morocco. Several other countries, including Egypt, Indonesia, Nepal, and several African countries are anticipated to join the program during 1996. Selection of countries is based on the relevance of participatory irrigation management to ongoing or planned World Bank loans and on the level of interest expressed by the host country and the concerned operational division.

For each country where it is offered, the program entails a series of three seminars (Phases 1-3) with the potential for further involvement in implementation (Phase 4) and evaluation/dissemination (Phase 5). The five phases are as follows::

Phase 1: A national seminar to introduce policy makers to the implications of PIM, consolidate national experience, learn about best practices from other countries, and formulate an indicative action plan for enhancing participation in the irrigation sector;

Phase 2: An international seminar held in a model country for PIM where participants can visit field sites to learn directly about the host country's experience and compare experiences with other participants;

Phase 3: A follow-up national seminar to formulate a national action plan for PIM;

Phase 4: Special-purpose seminars and/or training assistance to help support the implementation of a national PIM program; and

Phase 5: Support for evaluating national PIM programs and disseminating lessons learned to other countries.

Style and Methods

The seminars in phases 1-3 are designed to encourage the active participation and open discussion of all participants. Each seminar has the very practical goal of drafting an indicative or revised action plan for implementing participatory irrigation management. Presentations from international and national cases, and in particular the field visits, provide ideas to consider for including in the action plan. Much of the real work of the seminar is done in concurrent small group sessions where participants focus on a specific topic and discuss ideas in terms of their relevance to national conditions. The ideas that emerge from the small groups are presented to the plenary session for further discussion and possible incorporation into the indicative action plan.

Sponsorship and Cost-Sharing

Program activities are co-sponsored by WBI and one or more host country organizations, agencies, or ministries. The national co-sponsor normally covers all expenses for national participants (including air travel in the Part-II international seminar), with the understanding that funds can be drawn from ongoing World Bank loans. WBI generally covers the costs of international resource persons.

Irrigation Management: Benefits and Second Generation Problems

Participation in irrigation management by water users can take a wide variety of forms. Farmers can be involved in various system management functions, including, planning, design, operations, maintenance, rehabilitation, resource mobilization, and conflict resolution. Moreover they can be involved in these functions at various system levels; from the field channel to the entire system. Almost all irrigation systems have some involvement by water users in system management. When people speak of introducing "Participatory Irrigation Management" (PIM), they are thus usually referring to a change in the level, mode, or intensity of such participation that would increase farmer responsibility and authority in management processes.

Irrigation Management Transfer (IMT) is a more specialized term which refers to a process of shifting a number of basic irrigation management functions from a public agency to a private sector entity, a non-government organization (NGO), a local government, or to a local-level organization with farmers at its base. The most common form of IMT involves the shifting of management responsibility from a centralized government irrigation agency to a financially-autonomous local-level non-profit organization which is either controlled by the water users of the irrigation system or in which water users have a substantial voice in the control process.

The changes in management reported in the five primary case studies on which this paper is based can all be considered forms of IMT. However, there is an important difference between the organizational form of the recipient organization employed in the Philippines, on the one hand, and the form employed in the other four cases (Mexico, Turkey, Colombia, and Argentina). In the Philippines, the primary management unit employed is "community-based" and results from an intensive grassroots organizational campaign involving hired community organizers. This primary management unit is fairly small (less than 100 hectares), relies primarily on voluntary labour in carrying out its functions, and the most important relationships among members of the unit are social. In the other four cases, the organizational form of the irrigation systems can be termed an Irrigation Districts (ID). Irrigation Districts are typically larger (several thousand hectares), rely principally on paid employees to perform its functions, and link farmer-members together mainly through ties of economic self-interest.

Issues are analysed in this paper using a set of three basic categories:

(1) the processes used to introduce programs of IMT;

(2) the impacts of the introduction of a program of IMT; and

(3) second generation problems and possible solutions.

The term "second generation" requires some explanation. Transferring substantial management authority to a locally-based organization is a complicated undertaking and may involve changes in national policy, regulations, and organizational structures; creation of new organizations at the local level; transference of equipment ownership; and changes in personnel, in addition to the shifting of management functions to the new managers. Any undertaking this complex, in addition to solving problems, will almost certainly create new problems which did not exist before or were not previously evident. An example might be inadequate technical capability of new irrigation field personnel. These problems are here termed "second generation problems." In addition, there may be situations, such as low agricultural productivity, which were present prior to the transfer, but which were not acute problems when irrigation fees were low or non-existent. For our purposes, these are also included in the category of second generation problems. Some second generation problems may be a result of faulty processes used to introduce the new management system. Some may be a result of conscious choices during implementation to defer consideration of certain potential problems in the interests of accelerating program coverage. Others may be virtually unavoidable, though the ability to anticipate major problems in advance should allow corrective measures to be put into place earlier than would be otherwise possible. In this paper, problems are analysed from several different perspectives — that of the water user, the irrigation association, the irrigation agency, and the government. In terms of these perspectives, a change, such as increased irrigation service fees, may constitute a problem for water users, but a benefit for the irrigation association and perhaps the government, if it reduces government subsidies.

The term "impacts" also requires some explanation. In general, impacts can be either positive or negative. When they are positive, they are "benefits," or "positive benefits." When they are negative, they are similar to what are called second generation problems here. In the paper, negative impacts will be noted in the discussion of impacts, but will be discussed in more detail in the following section on second generation problems.

Introducing New Forms of PIM

Considering the case studies as well as PIM programs in other countries, key background conditions leading to the turnover process include:

- national budgetary crisis
- top level political will to place irrigated agriculture on a sound economic footing
- progressive deterioration of irrigation infrastructure due to deferred maintenance.

Only in one case (Colombia) of those reported at the workshop was it the farmers who initiated the process. At the irrigation systems level, another set of conditions that must be taken into consideration include:

- physical condition of hydraulic network;
- the social, political and economic conditions; and
- water availability.

Political will at the highest level of the government was a main component in Mexico's IMT program. In 1989, with a new administration in office, comprehensive water management was recognized as a top priority issue, the National Water Commission (CNA) was created and a national policy on privatization took off. In Colombia it was the National Planning Department who in 1991 submitted the Land Reclamation Program 1991-2000.

In Turkey, a budgetary crisis led to a squeeze on financial allocations to the state water resources agency (DSI) provided the initial impetus. In Argentina, a program to modernize the entire economy, which began in 1990 with the privatization of large electricity utilities, led to the turnover of water management to the provinces. In the Philippines the NIA embarked on an ambitious program in 1974 to increase rice production. A provision of this policy was for subsidies for the O&M of systems to be gradually phased out over a five year period at which time NIA would be directly dependent on collections from farmers for its O&M expenses.

The hydraulic infrastructure should be in fair condition and an affordable and reliable water supply should be available most of the time. Being in fair conditions means the hydraulic infrastructure is in operating conditions capable of delivering water to farms in sufficient amount to satisfy crop needs and in a timely manner. Surface drainage of surplus water and salinity should not be limiting factors. If these basic conditions are not satisfied, then a rehabilitation plan should be considered.

In Mexico, although the irrigation districts had suffered considerable deferred maintenance, the IMT program took off quickly because all of the systems were performing satisfactorily at the outset, with water conveyance efficiencies at main and secondary canal levels on the order of 60%.

The readiness of users to assume management responsibilities has to do with political and social factors. For example, it seems clear water users from Saldana and Coello in Colombia were "ready" in 1976 when they asked their government to turn over the administration of their districts to them. Due to its political background and social context it is apparent that Argentinean farmers were also ready for the change. In Turkey, farmers were used to a tradition of strong central government and many, especially the early adopters, were market oriented producers. This made them somewhat receptive to government declarations of the necessity of the transfer of managerial and financial responsibility.

In most districts of Mexico, however, this was not the case. In the first place there was the land tenure issue. After the 1910

Revolution land was divided between two types of small farmers, *ejidatarios*, who worked small plots of land held communally, and small landowners, some of whom belonged to the *hacendados* or landlords' elite. Second, there were numerous voracious, government controlled *ejido* leaders who, when consulted about the coming change, voted against it suspecting that the change would mean a reduction of their present status. Promotional aspects of IMT in Mexico country were vital in order to overcome these difficulties.

National Policies

National policies for IMT implementation vary both in regard to objectives and scope. Under the Philippines Water Code of 1976, the appropriation of waters by an irrigator association has priority over requests of individuals. The government then helped farmers organize themselves into irrigator associations which enter into various types of contracts with the NIA to handle O&M of the system.

In Turkey the General Directorate of State Hydraulics (DSI) is the main executive agency of the Government for the country's overall water resources planning, execution and operation. It was established in 1954 and is part of the Ministry of Energy and Natural Resources. Since the early 1960s, DSI has had a program to transfer O&M responsibility for secondary and tertiary distribution networks to IAs. Under the program IAs entered into contracts with DSI to take administrative responsibility for tasks such as collecting and submitting farmer water demand application forms to DSI, managing water distribution below the secondary canal, and cleaning and minor repair of canals and other small hydraulic structures. Although existing municipality law appears to be providing a workable initial basis for the formation of IAs, their further development and evolution may require a law specifically for IAs.

In Argentina the country is a federal entity divided into 23 provinces, which are autonomous in all aspects related to water (rights, granting, duration, taxation, etc.). IMT programs began in 1990 as a response to pressure from the national government to reduce bureaucracy and render public administration more efficient.

In Mexico the 1989 presidential decree that created CNA also granted the agency the responsibility to: (a) define the country's water policies and (b) allocate water to users through licenses and permits. The new policy:

- Created autonomous and self-financing and water utilities to provide water services in cities and in irrigation districts,

- Encouraged water reuse and water quality conservation, and
- Promoted a new water culture based on the efficient use of the resource.

In Colombia the policy of IMT is part of a larger shift in Agricultural Policy towards minimizing state subsidies and regulation. However the state continues to play a major role in land reclamation, and rehabilitation and expansion of irrigation areas.

Irrigation Service Fees

In the case of Mexico, the IMT program was part of a series of changes in the economy including reductions in subsidies for agricultural credit and inputs, elimination of guaranteed support prices for the major agricultural crops, and increases in energy and fuel prices. Transfer of O&M responsibility for the irrigation districts, leading to users paying the real cost of irrigation water, was seen as just another step in the liberalization of the economy.

In a similar fashion, in the case of the Philippines, a 1974 Presidential Decree authorized the National Irrigation Agency (NIA) to delegate partial or full management of irrigation systems to duly organized associations. Under this decree, NIA was allowed to keep all irrigation service fees, with government subsidies for O&M expenses gradually phased out over 5-years. Thus, at the end of this period NIA was to be directly dependent on ISF collections from farmers for O&M expenses.

The financial aspects of IMT in Turkey are similar to Mexico in that the policy is designed to shift the burden of O&M costs from the government to the users. However, the government continues to subsidize maintenance, which is not the case in Mexico and the Philippines. Colombia has also shifted the financial burden of O&M to the users, while irrigation schemes in Argentina are under joint management with fee collection by both the government and the IAs.

Water Laws

National water laws that clearly specify the rights of the IAs and the individual users appear to be an important factor in successful management transfer. Without such rights, the IAs are extremely vulnerable to increased demands from other more powerful users, such as industry and municipalities.

Beginning in 1976, the Philippines, attempted to develop a water rights register of all water rights in the country, including specifying in volumetric form the water rights of the national and communal irrigation systems. As part of the registration process all IAs must register their water rights. Once legally registered, water rights cannot be withdrawn except for failure to use them as stipulated in the law.

This system can be contrasted to that of Mexico where the IAs are given a limited concession to use the irrigation infrastructure and the associated water supply, but do not have a clearly specified legal right to a volumetric supply. With domestic supply having priority, the IAs are not ensured a constant water supply over time. The concessions are also for a fixed time frame (20-30 years) after which they can theoretically be reassigned to another user. Since none of the concessions have expired to date, it is unclear exactly what process will be used to determine granting of a second concession.

The country where IAs appear to be most vulnerable to changing water demands is Turkey, where water is controlled by DSI and the individual associations have no water rights. This system works in areas where there is little competition for water but leaves the associations extremely vulnerable in areas where municipal & industrial use is expanding rapidly. Colombia is somewhere in between these two situations but still is highly dependent upon the national irrigation agency, as the country has yet to establish a legal water rights system.

Mode of Implementing PIM

The process of implementing a change to a more participatory type management system varies widely from the bottom-up approach used in the Philippines to the more top-down approach used in Turkey and Mexico.

In promoting the PIM process, Colombia has invested much less time and effort than the other countries. As a consequence, the process of transfer was very quick but there have been second generation problems.

Users have been less well informed and have been uncertain about their rights with respect to ownership and changes in management practices. As a number of the irrigation systems in Colombia are based on river pumping, some of the more recent cases of transfer have involved additional negotiations with respect to energy subsidies.

Argentina has used information meetings and word-of-mouth to make a rapid transition to PIM, while Mexico and Turkey have used more structured informational meetings with users. These two countries also invested heavily in training their own staff. In Mexico, audio-visual materials prepared by IMTA and outside firms were used to persuade users that IMT was a positive change.

The Philippines has used the slowest process with intensive use of institutional development officers and farmer organizers (FOs) to serve as catalysts. These organizers lived in the villages and organized exchanges between NIA and the users. However, NIA has now realized that too much dependence upon FOs has slowed the process and has now reduced significantly the number of FOs.

In terms of transferring responsibility, most of the Colombia IAs appear to have had very little say in the process. However, this has changed somewhat as more recently the IAs have been negotiating energy subsidies and insisting upon rehabilitation prior to transfer.

In the Philippines there has been more dialogue through the organizers but given that only limited transfer of responsibility has taken place, this approach has not achieved rapid transfer of responsibility to the users. In contrast, both in Turkey and in Mexico there have been very active negotiations concerning transfer terms. As a result the IAs have been able to exert some power and develop a partnership with the agency based on a meeting of equals. In a number of cases they have also been able to ensure that critical investments were made prior to transfer.

As a key stage of the transfer process, Mexico, Colombia and Turkey have adopted a phase of shared management between the agency and the IAs during the transition. The duration of this shared management phase varies by country and by district but is usually between 6-12 months. In contrast, in Argentina the systems were transferred much more quickly, while in the Philippines there has been a very gradual shift in responsibility, in some cases over a period of many years.

Type/Nature of IA

The type and nature of the IAs are directly related to the structure of the economy as well as the type of irrigation found in the countries. In Mexico, Japan, Taiwan, Turkey, Argentina and Colombia, the IAs are larger (2,000-50,000 ha) and are organized more along the lines of commercial entities, reflecting the more commercial nature of the irrigated sector. Agriculture is developed on a cash basis and many of the staff are hired professionals paid in cash from the ISF. Given their large size, the IAs can afford to purchase and maintain their own transport and maintenance equipment.

In contrast, IAs in the Philippines are very small (100-300 ha) and are often organized based on the village structure. Most of the labour is voluntary labour provided by the users, and very few, if any, of the irrigation staff are hired professionals. Given the small size of the IAs there are diseconomies of scale and hence the organizations cannot afford to own specialized maintenance equipment. In the Philippines, irrigation service fees are usually paid in grain and therefore are very awkward to store and transport and typically result in 10-15% losses due to damage during storage and transport.

In Argentina, IAs are public NGOs with full legal authority, including the power to tax. The IAs in the other countries have a limited power to establish and collect ISFs but do not have any other local taxation powers and therefore the majority of their income is from ISFs.

Impacts and Benefits of Transfer

Implementing a program of management transfer is a complicated undertaking which involves incurring costs and affecting the lives and livelihoods of many people. It is thus not desirable to enter into such a program unless the benefits of the changes are positive and significant. Impacts, of course, may be either positive or negative, and they can be either qualitative or quantitative. And because the change in management patterns will usually occur simultaneously with other changes in physical, economic, and social conditions, it may be difficult to separate the effects caused by management changes from those caused by other factors. The nature of the impacts which occur will be shaped by the social, political, and economic characteristics of the countries involved. Impacts are also conditioned by the perspective from which they are viewed.

Important perspectives in this case include those of water users, the associations they have already created, the irrigation agency and the national or state government under whose overall control these systems operate. What is positive from one perspective might be negative for another. Judgment is thus required in evaluating the overall impact of a program, and the tradeoffs in positive and negative benefits among the various groups affected.

Legal and Organizational Impacts

Some of the countries first build on a new national water Law before moving into a transfer program (Mexico, the Philippines and

Colombia). Others move into transfer supported only by old national laws (Turkey) or local laws (Argentina). There has been a lot of discussion on the issue of water rights from the perspective of ownership versus concession. In most of the countries, all water belongs to the governments, either at federal or provincial levels. Irrigation Districts were usually created based on constitutional mandates that clearly defined the system's physical limits and thus the water right implicit to it. In some countries such as Argentina, however, the right of individual appropriation takes legal precedence.

After turnover, irrigation associations (IAs) in some countries receive the concessions which give them access to collective water rights. This does not change the existing system of water rights and priorities. In other countries, systems of water rights allocation and protection are virtually non-existent, which can lead to serious second-generation problems. In most of the case study countries, agricultural uses are subordinate to municipal uses, which include domestic and industrial use.

Sense of Ownership

The formation of IAs generally creates a stronger sense of ownership of the system on the part of users. In Japan, the Philippines, Colombia, and Argentina (Mendoza) this has been the case for many years. In Mexico and Turkey this change has taken place more recently and has been an exhilarating sense of pride for some users. Even though actual ownership of system facilities remains with the government. In Colombia some consideration is being given to transferring actual facilities ownership to IAs, an idea supported by some INAT staff and IA officers. This idea has also been discussed in the Philippines. There are a number of problems with this step, including loss of the right of eminent domain which is held by governments only and allows land condemnation for system expansion, and increased exposure to liability for accident and injury on the part of the IAs. Such a move may also weaken an IA's claim on public assistance for rehabilitation and emergency repairs.

A sense of ownership is also enhanced because users now have a greater voice in selecting the governance of their system, normally by electing boards of directors through the IA general assembly. With leaders elected in a democratic manner, needs can be expressed more readily. The opposite situation is presently occurring in Taiwan, where the users are turning systems back to the government. This is a result of the declining importance of agriculture in the Taiwanese economy and the high costs of system O&M which cannot be sustained by many IAs.

Transparency

All of the countries report that their transferred irrigation districts are now managed with more transparency, meaning by this that major decisions are exposed to the association's members or their representatives through board of directors and general assembly meetings. In Mexico, a representative assembly, formed by delegates selected by the general assembly takes routine decisions without summoning the general assembly, which may be composed of hundreds or even thousands of users. In Turkey a similar body made up of 30 to 70 local government officials and farmers also serves as a general assembly.

Transparency may be even more important in financial accounting than in organizational politics. Before transfer, the government collected fees and was generally supposed to invest the fees back into the system where they had been collected. However, due to bureaucratic procedures collected fees were not necessarily applied in the same systems where they originated. This resulted in complaints by users, and the agency was seldom forthcoming with information on financial accounts relative to particular systems. With associations collecting and managing funds, and with accounting done on the basis of the unit managed by the IA, users can get a better understanding of how their money is being used. However, strict supervision and auditing of water fee collection and expenditures by an outside party is still necessary to counter the possibility of money being diverted for unauthorized purposes.

Where the transferred units do not include an entire hydrologic unit, it may be more difficult for users to develop a sense of the disposition of the fees collected from them due to sub-division of responsibilities among a number of IAs. In this case, accounts should separate out the portion of fee collections that are earmarked for O&M within the IA unit, and the portion allocated to higher level O&M. An alternative might be for the IAs to merge into a larger unit. However there a number of other factors to consider before advocating such a step.

Water Fees Increase

Usually one of the immediate effects of IMT is an increase in water fees at the irrigation district level. In some countries the increase can be very dramatic, on the order of 100% to 200% or more (Mexico, Turkey). However to make a valid comparison, fee rates must be put in constant value terms before being compared. Irrigation service costs are generally considered appropriate if they constitute 5% to 8% of total production costs. In most countries, even after transfer, they tend to fall into this range. In the case of small holding sizes and lower value crops, however, the gross margin retained by farm households is also a relevant factor, and where this is already small, a doubling or tripling of ISF can create hardships for small farmers and their households.

A proposed increment in ISF may also have a negative psychological impact on water users that have come to expect heavily subsidized irrigation service. This kind of attitude may be gradually changed with promotion activities before transfer takes place and, of course, with efficient financial management and better service once the association is formed.

ISF collection rates are above 50% in all five countries currently implementing IMT activities, and are above 70% in four of the five. However, in looking at the share of O&M costs covered by ISF collections, the picture is a bit different. For example, before transfer in Mexico, the ISF collection rate was around 100%, but this amount covered only 25% of O&M costs. After transfer, the ISF collection rate is still 100% in transferred districts, but fees now cover close to 90% of O&M expenses. In order for transferred schemes to be financially self-reliant, they must achieve high rates of fee collection as well as having an ISF that covers O&M costs.

Subsidies

From the government's perspective, an increase in O&M cost recovery normally means a reduction of subsidies. By subsidy we define here any kind of costs associated with the provision of irrigation services that the farmers should be paying instead of the government. Expenditures within an ID are usually related to:

- O&M costs plus overheads,
- acquisition of premises,
- purchasing machinery and equipment,
- rehabilitation and/or modernization,
- technical assistance and training, and
- emergencies.

O&M plus overheads refer to expenses related to the daily administration of the system from the head works down through the tertiaries. These expenses have to be paid in full, either by the government, the users, or a combination of both if a system is to be sustainable. The sharing of costs between the government and IAs ranges from 40/60 to nearly 0/100, as is the case with Mexico.

Acquisition of premises refers to the land and buildings IAs use as they provide service to farmers. In Mexico CNA initially provided premises to IAs as part of the concessions. After six years of turnover though, some associations have purchased their own offices, which they often refer to with pride.

Providing machinery and equipment to IAs has been a key component of transfer programs in Colombia and Mexico. In other countries, like Turkey and the Philippines, machines are still under the control of the irrigation agency. In Mexico equipment provision was heavily subsidized by the government to encourage farmers to participate in the IMT program. Many of the machines given in concession to IAs were old and the government had to subsidize repairs. Other machines were purchased under a World Bank loan. As lowering of maintenance costs was a concern for both government and users, emphasis was put on the purchase of light machines to substitute for the heavy construction equipment which had traditionally been used in the ID by CNA.

Rehabilitation and/or modernization of hydraulic infrastructure is also a key issue for all countries involved in IMT programs. Although farmers usually ask for investments of this kind to be made before turnover takes place, governments are often unwilling to supply the necessary resources to fulfil these requests. Turkey is an exception to this rule, where rehabilitation was not an integral part of the IMT program. In some cases countries use international loan funds for this purpose, enabling the government to negotiate with farmers the amount of needed investment and/or the financial share that users will need to contribute. Costs are often shared between the government and IAs on a 50/50 basis (e.g., Mexico).

Support for technical assistance and training may also be subsidized through IMT programs. In Mexico, technical training for both IAs and government irrigation district staff, has been provided by CNA with assistance from the Mexican Institute for Water Technology (IMTA). More recently, starting in 1995, some IAs have signed private agreements with IMTA in which technical assistance is being provided on specific projects with a limited subsidy from the CNA. In the Philippines, promotional and technical support to farmers is regularly provided by the government through various types of community organizers. In Turkey, extensive training for agency field staff and farmer leaders was provided by the irrigation agency. Yet, in all countries the training needs for PIM programs by far exceeds current investments in training.

Subsidies are also normally available in the case of emergencies situations. Although little specific information on this is available, all of the countries recognize the need of governmental intervention when such cases do arise, e.g., hurricanes in Mexico or typhoons in the Philippines. Water scarcity also creates emergency situations for transferred IDs. For example, during the 1995-96 irrigation season, an extraordinary drought took place in northern Mexico. Due to the lack of water, IAs did not have water to deliver to farmers; consequently, and collection of fees dropped nearly to zero. For many of the IAs the government had to step in with 100% subsidy programs to avoid a collapse of the associations. A few IAs that were only partially affected by the drought tackled the problem by hastily imposing a compensation fee on their farmers to implement a water reuse program.

Impacts on Operational Procedures

It is expected that a shift to participatory local management will improve the effectiveness and efficiency of water delivery as well as the quality of maintenance. The case studies provided some details to document these impacts but there is still a need to obtain better information about the operational impacts of IMT.

Improved Maintenance Programs

One of the immediate consequences of IMT programs is that irrigation systems have improved maintenance programs. (Colombia, as recently reported, seems to be an exception). A recent study carried out in 1994 in Mexico by the Colegio de Post-Graduados reported that 84% of the sample users believed maintenance had improved substantially since IAs have been in charge of O&M. In some areas not only have IAs completely eliminated the problem of deferred maintenance, but are also are now putting surplus money from O&M activities into programs for the modernization of hydraulic infrastructure.

Improved Services

Although few specific studies on quality of services after IMT exist, the general impression is that, after turnover, services have substantially improved in regard to timeliness, reliability, and equity of water distribution. There are many reasons to believe this. In the first place, after turnover users usually have better access to the irrigation service provider. Compared with the relationship with the agency that existed before, the farmer-association relationship is much better in terms of distance, personal contact and feedback on complaints. Water conflicts among farmers also tend to be minimized because farmers usually search for solutions to their problems at the IA level which is physically closer to them. There is also a degree of self-control prevalent, as the users know each other very well and are therefore able to regulate behaviour among themselves.

In a 1994 study of IAs in Mexico, 84% of the users indicated that water distribution had improved since turnover, 79% said they were receiving enough water, 79% were also receiving water in time, and 64% indicated water was being allocated in the appropriate amount. However, water conflicts do arise between the IDs and IAs, especially when water is scarce. Recent lessons from the Mexican drought indicate that, under extreme competition for water, IAs were overwhelmed by operational problems. This required prompt government intervention to keep the conflicts from getting out of control and becoming an explosive political issue.

Agency Re-organization

IMT programs can have a significant impact on agency organization. A sharp reduction in field personnel and O&M staff usually takes place. A change in the role of the agency with regard to water management in the systems follows.

Staffing: Once the government's employees perceive the possibility of losing their jobs as a consequence of IMT implementation, they may attempt to block the turnover process. Clear legal arrangements between the parties are needed in order to overcome difficulties. Unfortunately, as the number of people involved in the process may be quite large. In Mexico, 5,000 out of 7,000 government personnel were to be released from the IDs because of turnover, while the Philippines has reduced government agency staff from 19,353 to 10,368.

Changing the role of the government irrigation agency is a natural outgrowth of IMT. The change requires not only willingness from the part of high-level officials but also training for them on specific issues. One is simply convincing governments officials that there have been substantial changes in the way the irrigation sector is structures. For instance, in some countries the agency keeps asking from IAs the same types of information as if management had never been transferred. In Mexico, after six years of turnover, changes in the governmental sector are still incomplete.

Once the turnover thoroughly takes place in a system, the agency's new role should be more oriented towards supervision, guidance, monitoring and regulation of water rights, plus selected technical assistance support to IAs. Management responsibilities of the system should be in the hands of the users, with the possible exception of head works which when considered of strategic importance or technically sophisticated, should remain under agency's direct control and administration.

Second Generation Problems and Solutions

As noted in the introduction, problems and solutions in a particular location will differ, depending on the perspective of the observer. The discussion below addresses problems from the perspectives of (a) the irrigation association, (b) farmers, (c) the irrigation agency, and (d) the government. Solving these problems can involve a variety of steps including revision of laws or implementing regulations and changes in organizational structure, organizational rules and processes, and funding mechanisms.

In addition, associations will require supporting services. Support services are services which come from outside the association itself but which are necessary for it to carry out its mission. They include such things as financing for equipment purchases, legal advice, computer programming assistance, and financial auditing services. Such services may be difficult for an association to generate internally (e.g., financing for heavy equipment) or be used only infrequently, (e.g., specialized maintenance equipment) and hence be too expensive for an association to maintain on a full-time basis.

Support services can be procured by associations from a variety of sources—private firms, public agencies, universities and institutes, non-government organizations, and regional or national federations of associations. In the past, it has usually been assumed that any such services must come from government agencies. Today it is recognized that higher quality and less expensive services may often be obtained from other sources, and that the government should generally serve as only one of several alternatives, rather than the sole source.

Irrigation Associations

Insecurity of water rights was identified as the most serious second generation problem affecting irrigation associations in the five case study countries. Water rights which are often absent, poorly-defined, or insecure, can (a) inhibit investment in new system facilities or rehabilitation, (b) encourage short-term thinking and behaviour on the part of association managers and farmers, (c) result in heavy expenditures in legal costs to defend a poorly-defined water right, and, ultimately, (d) lead to a reduction in water supply and even system collapse.

An effective water right should provide security to the association, but must, at the same time, be adaptable so that water can be diverted to other more productive or higher priority uses as economic and demographic conditions change. In this event, there must be a provision for appropriate compensation to those who are giving up the water right by those who gain it. In Mexico, for example, an association's irrigation water right is always subservient to present and future municipal demands. This creates considerable insecurity for associations which share water sources with growing municipalities and violates both the principles of security and that of just compensation.

An effective water right should be specified in both quantitative and qualitative terms. Water quality degradation by upstream effluent discharges, as from a factory or an inadequate municipal sewage treatment plant, can render the water unusable by downstream irrigators. It can also make the water suitable only for lower value crops, since biologically or chemically contaminated water may not be permitted for production of higher value fresh fruits and vegetables. This will become an increasingly serious problem as water reuse increases in response to growing demand from all sectors.

Establishing a water rights system where it is lacking, as in Turkey, or clarifying water rights where they are weak, ineffective, or inequitable, as in Mexico, will usually require action from the national legislative body or from top level executive leadership, or both. It is thus extremely important for water user associations to have adequate representation of their interests when these issues are taken up. Two different types of support services are identified as being crucial for associations in attempting to establish or clarify their water rights. The first of these is legal advice and representation when the association faces challenges to its rights. Such representation is best secured from private law firms, if available, since services secured from government sources may be of lower quality, and may be subject to political pressures which would compromise their objectivity.

Legal representation is also important during the formative stage of an association, when negotiations with the government irrigation agency will establish the contract or concession which will control the relationship of the association and the government. Unfortunately, associations which are just forming may be unaware of the importance of high quality legal advice at this stage or may be unable to afford it. A national federation of associations can play a valuable role as a source of legal advice and assistance to newly forming associations.

The other type of support service required by associations is lobbying on their behalf in government policy making councils. Since other interests, such as municipalities and industrial water users, are usually larger than individual associations and likely to be more powerful politically, it is important for associations to establish regional or national federations representing many associations and a large number of farmers. This will give them political influence with which to counter the lower of competing interests. A federation would also be to represent irrigation associations on the board of directors of the national irrigation agency as is currently the case in Colombia.

Financial shortfalls comprise another high priority second generation problem affecting associations. A central feature of the PIM programs undertaken in all of the case study countries is financial autonomy. Financial autonomy is the condition where an organization generates all of the revenue it needs to support itself and to perform it's primary functions.

It implies that the association is not directly subsidized by the government, or that if it is subsidized, that the subsidy is a fixed amount which does not vary according to the condition of the association's balance sheet. The principal source of revenue for most associations is irrigation service fee (ISF) collections. Financial shortfalls affect a number of associations in Colombia and Argentina have been a concern in Mexico since the 1995 economic crisis. Financial shortfalls are a function of several factors, including ISF rates, ISF collection effectiveness, the contribution of other sources of revenue, and expenditure patterns. One important factor is the structure of the ISF. Fees can be levied on flat rate or volumetric bases. A recommended structure for fees is a two-part one comprising both fixed and volumetric components.

The flat portion would constitute a "connection charge", a charge for simply being within the boundaries of the system's service area whether or not water was actually taken from the system. This would reimburse the association for expenses incurred in maintaining the physical and administrative capacity to deliver water to the farm. The absence of this component in the fee structure of Mexican associations has created severe problems during years when drought greatly reduced the available water supply to the system's water users. The other portion of the fee would be based on the volume of water actually delivered during a cropping season, or some proxy for this amount, such as area irrigated and number of irrigations given. This would cover the costs incurred by the association which are related to the amount of water given, and would serve to limit excessive demand for water. Revenue from ISF is also dependent on the percentage of the fees assessed which are actually collected, though associations in many countries do a reasonably good job in this regard. An exception is the Philippines where low collection rates have been a persistent problem for NIA.

Solving problems of revenue shortfalls that relate to fee levels and collection efficiency is largely an internal association responsibility. Outside assistance may be useful, in some cases, in estimating farmers ability to pay particular ISF levels, and in analysing management systems set up for collecting revenue. This is discussed further in a following section. Underlying difficulties in generating sufficient ISF revenue to sustain system operations, in many cases, is the low productivity of irrigated agriculture in system command areas. Low productivity can result from a large number of factors, but is often associated with small farm size, a subsistence orientation, production of low value crops such as grains, inappropriate agricultural policies, a poor natural resource base, and inadequate agricultural support services.

In such cases, a solution to the association's financial problems may be possible only if the underlying problems in the agricultural sector are addressed. If these problems cannot be solved, then the options are for the government (a) to have other agencies provide technical assistance to increase production, or (b) provide the association with special subsidies. As a last resort, the government may have to consider taking back the responsibility of system management and financing. However, as irrigation service fees typically constitute only 3 to 10 percent of total production costs, reducing them will generally not solve underlying problems of high agricultural production costs and low productivity.

Rehabilitation is identified as a third important second-generation problem for irrigation associations. All irrigation systems require periodic rehabilitation and modernization. While usually less expensive, in real terms, than the original construction, rehabilitation is a costly undertaking, and is usually beyond the financial and technical means of an association to undertake. Seldom is there a clear and consistent government policy on responsibility for rehabilitation. In the absence of such a policy, the tendency is for associations to defer needed rehabilitation in the hope that the government will step in and take responsibility for it. In this case, IAs usually underinvest in system improvements between rehabilitations. This tendency is reinforced by the government retaining ownership of system physical facilities, while transferring to associations only the use rights of the facilities. Associations may thus regard the responsibility to rehabilitate those facilities as belonging to the government unless a different policy is clearly stated in the agreement between the government and the association.

A related problem is that of cost sharing between the government and the association for rehabilitation. Because irrigated agriculture benefits people beyond the ranks of system irrigators, and because full coverage of rehabilitation costs is usually beyond the means of the irrigators themselves, a sharing of costs is appropriate. Responsibility for even a share of the costs involved will tend to counteract the tendency of an association to defer maintenance, as noted above.

To cover its share of future rehabilitation costs, associations usually need to accumulate a capital replacement fund over a number of years and there needs to be a legal basis for establishing such a fund. There should also be incentives for establishing and contributing to such a fund. Unfortunately, fiscal and monetary policies in many countries such as Turkey and Mexico, have led to high rates of inflation and low or negative real interest rates on savings, which acts as a powerful deterrent to fund accumulation.

Governments may need to create special investment opportunities for associations which allow them to earn reasonable rates of return on accumulated funds. Likewise, there should be incentives for associations to make improvements in physical infrastructure. One way to do this is to establish a trust fund, perhaps with donor financing, from which associations could request funds to complement their own investment funds.

The matching ratio for such a funding facility should be established and made known in advance. A number of support services are required specifically to support system rehabilitation. These include

(1) assessments of system facilities, (2) credit, and (3) design and construction assistance. Regular assessments of the condition of system facilities can be done jointly by the association and the government agency, as in Turkey, or may be contracted out to an engineering consulting firm acceptable to both the association and the agency. Such assessments can be used as a basis for annual maintenance planning, to suggest the need for selective improvements in system facilities, and for planning whole-system rehabilitation.

If an association is unable to accumulate its share of rehabilitation cost prior to rehabilitation and does not have the ability to assess a special charge on the members, it will need a source of credit. Credit can come from private banks, government banks or other lending facilities, or from insurance pools in the case of rehabilitation induced by natural disasters such as floods, typhoons, or volcanic eruptions. Such a credit facility could also be used as a source of financing for capital equipment needed for system maintenance.

Rehabilitation will also require external technical services for design and construction. Because of the sharing of costs, both the association and the government should be involved in decisionmaking relating to the selection of consultants and contractors and monitoring their performance. Advice and guidance to the association on handing these tasks might usefully be given by a federation of associations, since rehabilitation occurs only infrequently in any one association.

Lack of financial and administrative management expertise is the final topic identified as a high-priority second generation problem by workshop participants. There are several possible responses to this problem. One would be skill enhancement through staff training programs. Skills can also be enhanced by replacing less skilled people with more capable ones. Contracting out for specialized services is another important way of addressing management deficiencies in associations.

One extremely important step in improving the quality of association management is to increase the transparency of management processes. This has a number of positive effects. It can (a) reduce the potential for misappropriation of funds, (b) help insure that salary levels and benefits are realistic, (c) help insure that maintenance allocations are appropriately targeted, (d) reduce favouritism in making personnel appointments, and (e) improve responsiveness of association staff to users. A number of steps can be taken to increase transparency in association management. These include:

- regular external audits of financial accounts,
- use of standardized budgeting and accounting frameworks,
- wide dissemination of simplified budgets, plans, and financial statements,
- active involvement of the board of directors in forward planning, budgeting, and auditing, and
- broad representation from among users on the association board of directors.

There is broad scope for employing external support services to improve management of the association. Services that may be required include:

- advice on establishing and revising management systems and procedures,
- advice on establishing financial budgeting and accounting systems, including software,
- establishment of standard budgeting and accounting formats,
- standards and requirements for regular external audits, and
- management training.

These services can be obtained by the association from a variety of sources, including private firms, a national or regional federation of associations, NGOs, government agencies, and universities and training institutes. The government is the logical party to establish mandatory standards and requirements for external audits, but the audits themselves, could be done by a private firm of chartered accountants. Other services, such as management and accounting system advice could come from a variety of sources, with private sources being generally preferred.

One argument in favour of the provision of these services by government agencies will often be that they can be obtained at no or low cost. What makes this lower cost provision possible, however, will generally be implicit government subsidies to the service providers. A preferred alternative would be to provide the funds supporting these subsidies instead to the associations as grants to be used for obtaining management support services. This would allow the associations to contract for these services among alternative providers. The demand-driven competition thus induced would be a very healthy force acting to hold down service prices and improve quality of services delivered. Provision of such grants during the transition phase from government agency to association management could be a very useful institution-strengthening activity.

Irrigation Agencies

Dislocation of staff is the most prominent problem experienced by agencies following irrigation management transfer to users. This problem is typically dealt with in several ways.

First, O&M staff levels are reduced by attrition. When positions become vacant due to retirement or resignation, they are left vacant or filled by internal transfers rather than new hiring.

Second, financial incentives are often provided for early retirement of older staff. Third, existing staff are transferred to other positions which become vacant rather than filling them from outside the agency. In some countries such as China, where it is difficult to lay off staff, sideline enterprises have been created which can generate income for the irrigation district and cover the salary costs of the involved personnel. In the Philippines, this has taken the form of an irrigation consulting company, NIA Consult, which was created as a subsidiary of the National Irrigation Administration to provide irrigation advisory and design services within the Philippines and abroad.

In some cases, redundant agency staff may also be re-employed by the newly created associations which take up management responsibility for the schemes. Such employment should be at the discretion of the association. Loss of technical capacity in the national irrigation agency is a common problem faced by agencies which transfer management significant irrigated areas to associations and experience loss of technical staff. To address this problem, agencies can

- obtain specialized expertise from outside consulting firms as needed,
- increase salaries to attract and retain high-quality staff,
- provide in-service training opportunities for staff, and
- revise job descriptions to bring in new staff with the desired qualifications.

Defining a new role for the agency is another important challenge. With their operational responsibilities transferred to associations, the agencies need to design a new role to address emerging problems. Doing this requires discussion among staff within the agency and also at higher levels of the government, with broad participation by all involved parties. The aim should be to build broad consensus and political commitment for the new role. In some cases, changes in legislation may be required to enable the assumption of new responsibilities.

The new mandate should contain a clear definition of roles and responsibilities and should define skill requirements to carry out the new responsibilities. It should also contain a timetable for accomplishing the shift to the new mandate. Support services which could be useful in this process include:

- comprehensive diagnosis of the agency/association relationship and the associations' support needs,
- professional assistance with the agency's strategic planning process, and
- consulting services to design new management information systems for the agency.

In Colombia, the national agency, INAT, is employing professional consultants to help them define a new role for themselves under an Inter-American Development Bank credit.

Farmers

Second generation problems experienced by individual farmers relate mainly to the need to increase farm productivity to pay higher irrigation fees and to take advantage of possible improvements in irrigation service quality. Support services required may include:

- production credit,
- extension advice,
- new technologies,
- markets and market information,
- access to inputs, and
- post-harvest services.

Although government agencies are the traditional source of many of these services, in many countries, private or other organizations are playing an expanding role in supplying some or all of the services listed above. There is also the question of the potential role of the association itself in providing other agricultural services, in addition to irrigation service. As a general rule, the association should demonstrate competence in its core activity of irrigation management before considering such ancillary activities as providing other agricultural inputs.

Government

The principal second-generation problem for government, beyond those already identified for the irrigation agency, is the reduced control which it will have over irrigation activities at the system level, and a diminished ability to use irrigation as a tool to implement other national policies and priorities. An example might be the government's wish to promote cultivation of upland crops rather than rice during a particular season.

In the past it could work through the national irrigation agency to adjust water delivery schedules and volumes to try to achieve this end. Following transfer, this becomes more difficult. There are other tools, such as support prices and subsidies, to achieve the same ends, however, so that losing control of irrigation system management should not pose a significant problem for agricultural policymakers.

Experience is now available from a number of developing countries which have recently implemented Participatory Irrigation Management (PIM) programs and there is additional evidence from developed countries which transferred irrigation management functions to locally-based organizations many years ago. In the case of the developing countries, while the overall benefits of PIM have been positive, in some cases second generation problems have manifested themselves and, consequently, have tended to reduce the magnitude of the potential benefits.

In the process of introducing PIM programs, political will at the highest level is a critical background condition for a rapid and sustained transfer program. A second important element is that the irrigation infrastructure be in fair condition so that it could deliver irrigation water as required. A suitable legal framework is also necessary for the sustainable functioning of the transferred systems. Lack of clear water rights has led to second generation problems including conflicts with municipal and industrial users as well as other irrigation organizations.

PIM is designed to shift the financial burden for irrigation service from the agency to the users. This aspect has to be made very clear when the process of transfer is introduced. Failure to address the financial side of system management is a primary cause of second generation problems. In general, the countries that have a clear policy on irrigation service fee rates and collection practices have sustainable water user associations.

The type and nature of the associations are very dependent on the structure of the broader economy as well as the type of irrigation and the tradition of management prevailing in the country. Where economies are more developed and diversified and irrigation systems are large, associations have tended to evolve successfully. These associations are generally large and can hire their own staff and own specialized irrigation equipment. In contrast, in countries where economies are less developed, agriculture is more subsistence oriented, and irrigation systems are small, the associations tend to be small and more problematic in terms of management.

These changes in management structures and processes have had important impacts both positive and negative on four important target groups: farmers, the irrigation association, the irrigation agency, and the government. For example, while increased service fees have reduced the financial burden on the government and increase the sustainability of the IAs, they have added to the costs of production for the users. On the other hand, from the perspective of the farmers, transfer has resulted in a sense of ownership, reduction in conflict and improved maintenance.

Transfer has reduced the O&M staff of the irrigation agencies, and consequently the overall number of civil servants working in the irrigated agriculture sector. However, in a number of cases, this has also reduced governments' control over cropping patterns and over water resources more generally. This reduced government control has generally led to greater farmer satisfaction, more transparency in decisionmaking, and greater overall economic efficiency.

Changes in management responsibility have led to second generation problems in most countries, some of which are already affecting the involved parties while others loom as potential future problems. Insecurity of water rights was identified as the most serious second generation problem affecting transferred systems. The primary solution is to establish a secure legal right that has both quantitative and qualitative dimensions. Federations of associations can provide assistance and legal representation.

Financial shortfalls comprise another second generation problem. The principle source of revenue for the association is the irrigation service fee. A two-part fee consisting of a fixed connection charge and a volumetric charge can provide greater financial stability. Increasing the productivity of irrigated agriculture is also an important element in improving the financial health of associations. Outside assistance may be useful in analysing and improving management systems set up to assess and collect service fees.

Rehabilitation poses a number of second generation problems. In the absence of a clear and consistent policy on rehabilitation, maintenance is often deferred. There is a valid argument for developing a cost sharing formula where the government pays a share and the users pay the remaining share. If possible, the IAs should have a sinking fund for accumulating funds to cover their share of future rehabilitation costs. When this is insufficient, it is important to have an outside source of affordable credit. Other needed supporting services include assistance with maintenance assessment practices and technical design and construction services.

Effective financial and administrative management of the associations requires specialized staff training and increased transparency. Support services such as external audits, and standardized accounting packages can also contribute to better management of the associations.

Irrigation agencies also suffer from a number of second generation problems. Dislocation of staff, loss of technical capability, and the need to define a new role for the agency are all problems found in countries that are instituting PIM programs. In particular, the problem of what to do with excess staff is a difficulty faced by almost all agencies. Solutions include attrition, retirement incentives, creation of specialized consulting units, retraining and assistance with job placement, and transfer to other units. Along with the problem of staff displacement, agencies also face the problem of the loss of specialized skills. These may be replaced by outside private consultants but may also require the agency to provide specialized training for existing staff.

Second generation problems of farmers are related to the need to increase agricultural productivity, including the need to shift to higher value crops. Services such as credit, agricultural extension, market access, technical inputs and post-harvest assistance are all needed. In some cases these services will come from the government but increasingly from the private sector as well. Federations of IAs can also play an important role in the provision of such services.

A shift from public agency control to local participatory management is unlikely to happen without some second generation problems. Rapid institutional change will almost always require corrective measures to address unexpected problems. Countries with flexible policies and procedures will be able to address these problems, as they arise. This report has summarized a number of solutions countries have employed to address second generation problems associated with the shift to participatory irrigation management.

5

Recent Advances in Dryland Agriculture

Indian agriculture is predominantly a rainfed agriculture under which both dry farming and dry land agriculture is included. Dry faring was the earlier concept for which amount of rainfall (less then 500 mm annually) remained the deciding factor for more then 50 years. In modern concept, dry land areas are those where the balance of moisture is always on the deficit side. In other words, annual evapotranspiration exceeds precipitation. In dry land agriculture, there is no consideration of amount of rainfall. It may appear quiet strange to a layman that even those areas which receive 1100 mm or more rainfall annually fall in the category of dry land agriculture under this concept. To be more specific, the average annual rainfall of Varanasi is around 1100 mm and the annual potential evapotranspiration is 1500 mm. thus the average moisture deficit so created comes to 400 mm. this deficit in moisture is bound to affect the crop production under dry land situation ultimately resulting into total or partial failure of the crops. Accordingly the production is either low or extremely uncertain and unstable which are the real problems of dry land in India.

The success of crop production in these areas depends on the amount and distribution of rainfall, as these influences the stored soil moisture and moisture used by crops. The amount of water used by the crop and stored in the soil is governed by the water balance equation: $ET = P-(R+S)$. When the balance of the equation shifts towards right, precipitation (P) is higher then ET, so that there may be water logging or it may even lead to run off (R) and flooding. On the other hand, if the balance shifts to the left, ET becomes higher then the precipitation, resulting in drought in the various severity. Taking the country as a whole, as per meteorological report, severe drought as large area is experienced once in 50 years and partial drought in five years while folds are expected every year in one part of the country or the other, especially during rainy season. In fact the balance of the equation is controlled by the weather, season, crops and cropping pattern.

Out of 14.2 million ha of net sown area in the country, rainfed agriculture is practiced in 95 million ha (67%). Nearly 67 m ha of rainfed area falls in the mean annual precipitation range of 500-1500 mm. The average annual rainfall of the country is 1200 mm amounting to 400 million ha meter of rain water over the country's geographical area (329 m ha). However, the distribution across the country varies from less than 100 mm in extreme arid areas of western Rajasthan to greater than 3600 mm in NE states and 1100 mm from east coast 2500-3000 mm in the west coast. The broad area of the summer monsoon activity extends between 30^0 N to 30^0 S and from 30^0 W to 16.5^0 E. the detail information on rain fall and monsoonal pattern in India has been summarized in the following table:

Table: *Rainfall pattern in fall*

Season/Period	***m ha m***	***Percent***
Winter (Jan-Feb)	12	3
Pre-monsoon (Mar-May)	52	13
Southwest monsoon (Jun-Sept)	296	74
Northeast monsoon (Oct-Dec)	40	10
Total for the year	400	100

Rainfed farming comprises about 91% area of coarse cereals (sorghum, pearl millet, maize and finger millet), 91% pulses (chickpea and pigeon pea), 80% of oilseeds (groundnut, rape seed, mustard and soybeen), and 65% of cotton. Also, about 50% area under rice and 19% area under wheat is rainfed.

During the past 25 years there occurred significant changes in the area and yield of imported crops of rainfed farming areas. The area under coarse cereals decreased by about 10.7 million ha and most of this was under sorghum. The area under oilseeds increased by 9.2 million ha and most of this increase was due to irrigated rapeseed and mustard and soybeen. The total area under pulses and cotton remained constant but more of cotton became irrigated and shifts in the area under occurred from one agro-ecological region to others. Area under chickpea in the northern belt decreased but increase in the central belt. This change occurred due to increase in area under rice-wheat cropping system which displaced chickpea and also pearl millet to a great extent and maize to a small extent.

According to the present concept, there are 128 districts in the country which face the problems of dry land of these 25 districts covering 18 m. ha of net area sown with 10 % irrigation receive 375-750 mm rainfall annually spread over Central Rajasthan, Saurashtra region of Gujarat and rain shadow region of Western Ghats in Maharashtra and Karnataka. Twelve districts have irrigation covering 30-50% of the cropped area and do not pose serious problems. The remaining 91 districts covering mainly *Madhya Pradesh, Gujarat, Maharashtra, Andhra Pradesh, Karnataka, Uttar Pradesh, parts of Haryana, Tamil Nadu* etc., represent typical dry land area. The total net sown area in these districts is estimated to be 42 million hectares of which 5 m ha are irrigated. Rainfall in these districts varies from 375 to 1125 mm. therefore, more and more efforts are to be made for enhanced and stable production in these areas so that the recurring droughts do not stand in the way of meeting the growing food demands.

It is not that no attention has been paid in the country towards the development of dryland farming. Efforts were made right from 1923 to improve crop yields with the establishment of a research projects at Manjari in Maharashtra and later at Solapur, Bijapur, Raichur and Hagari in Deccan and Rohtak (Haryana) in the north. An All India Coordinated Research Project for Dry land Agriculture was launched by ICAR in 1970 in collaboration with Government of Canada and later Central Research Institute for Dry land Agriculture (CRIDA) was established in 1985 at Hyderabad. These projects generated technology, which, if followed, can bring marked improvement in cropping intensity, productivity and stability in production.

In dry land agriculture, scarcity of water is the main problem. Apart from the low and erratic behaviour of rainfall, high evaporative demand and limited water holding capacity of the soil constitute the principle constraint in the crop production in dry land area. Yield fluctuations are high mainly due to vagaries of weather, often much behind the risk bearing capacity of the farmers. It is surprising to a layman that even humid areas with 2000 mm of annual rainfall not only suffer from moisture stress, but also face drinking water scarcity. Monsoon starts in the month of June and ends in last week of September or sometimes in the first week of October. Most of the rainfall is received during this period. With undulating topography and low moisture retention capacity of the soil, major portion of the rain water is lost through runoff, causing erosion and adding to the water logging of low lying areas. After the rain stops, very little moisture is left in the profile to support plant growth and grain production.

In dry land area deficiency and uncertainty in rainfall of high intensity causes excessive loss of soil through erosion which leaves the soil infertile. Owing to erratic behaviour and improper distribution of rainfall, agriculture is risky, farmers lack resources, tools become inefficient and ultimately productivity is low.

Vertisoles have high clay content and high moisture retention capacity. Owing to its swelling and shrinking characteristics, permeability is low and hence the rate of infilteration of water is minimum. This causes more surface and high soil loss from the top layer owing to surface erosion. It is estimated that 68.5 tones/ha per year soil is lost from *vertisoles*. Due to high clay content it develops cracks during Rabi season at flowering stage of crops.

Alfisols are, by and large, light textured soils which have low moisture holding capacity but high water intake. The rain water falling in such areas gets soaked up and saturates the profile. The soil water percolation is more and therefore, is lost for crop use. Owing to faster intake of water in the profile the surface runoff is limited and soil loss from erosion is low (3.05 t/ha/year). Soil crusting is a common problem in low rainfall areas.

Entisols are generally loamy sand or sandy loam. Depth in these soils is not a constraint. These soils have very low clay content and hold water up to 200 mm per meter of soil profile. Its nutrient holding capacity is poor. In low rainfall areas monsoon cropping is practiced and in high rainfall areas double cropping is possible.

Submontane soils are medium in texture and depth is medium to deep as well as moderate in clay content. Moisture retention capacity is high (300 mm/m. profile). These soils are poor in nitrogen but in other nutrients. Phosphorous may be limiting in high production system. Due to high rainfall double cropping is possible in these soils.

Sierozems are extremely light soils, effectively depth being influenced by the $CaCo_3$ concentration in soil profile. Its moisture holding capacity is low (150 mm water.m). Sierozemic soils are low in nitrogen and sometimes inadequate in phosphorous. Subsoil salinity is common. These soils are mostly monsoon cropped, except in deep sandy loams where post-monsoon cropping is also possible. Crusting is very frequent.

Improved Dryland Technology

The improved techniques and practices, which have so far been generated and recommended for achieving the objective of increased and stable crop production in dryland areas, have been summarized in following lines.

Crop Planning

The farmers of the dry land areas, prior to the development of dry land techniques, were growing a crop either on rainwater in kharif or on conserved soil moisture during the winter. The crop varieties grown when moisture is sufficiently available. Such varieties have low genetic potential for yield. Selecting suitable crops and varieties capable of maturing with in actual rainfall periods will not only help in enhancing production of a single crop but in intensifying the cropping intensity. Many criteria have been laid out for selecting a crop variety for drylands. The capacity to produce a fairly good yield under limited soil moisture conditions is the most desirable criteria.

The duration of kharif crops/varieties should not normally exceed the number of rainy days. In other words, crop varieties for dryland areas should be of short duration, through resistant tolerant and high yielding which can be harvested with in rainfall periods and have sufficient residual moisture in soil profile for post-monsoon cropping.

Under dry land agriculture determination of length of growing period (LPG) i.e., moisture availability of a given soil type, provides better index than total rainfall based crop planning. LPG is defined as the period when the moisture and temperature regious are suitable for crop growth and the period is determined by the FAO method (1983). The LPG is computed as the sum of the period when P is more than 0.5 PET plus time taken to utilize stored soil moisture (assured 100 mm) after P falls short of PET. For example 'Nagpur' and Ratnagiri in Maharashtra receive mean annual rainfall of 1120 mm and 2500 mm, respectively but LPG determination indicates that both the places have LPG of 210 days in deep black soils. Therefore, both the places are suitable for single long duration on a short duration crop with a relay rabi crop.

Planning for Aberrant Weather

Dryland agriculture is subject to high variability in areas sown, yields and output. These variations are the results of aberrations in weather conditions, especially rainfall. Delays in normal monsoonal pattern causes problems of timing and the organization of preparatory tillage and other initial activities for commencing cultivation processes for the season. Such monsoonal delays have repercussions on the programme of activities for the entire agricultural year. Even after the onset of monsoon and the commencement of planting, there may be monsoonal withdrawal causing moisture stress on plants and creating difficulties in the adoption and timing of approval cultural practices ultimately causing reactions in yields and outputs. Some crops are highly susceptible to such mid-season variations in moisture availability such as at the flowering stage in rice. Major crops like rice and maize get seriously affected if monsoonal rains cease early.

The need for modifying and introducing and introducing new technology for increasing and sustaining yield in dry land areas can hardly be overemphasized. Equally urgent is the need to decelerate and ultimately eliminate the process of damage to agricultural assets which are proceeding unbated in dry land areas. Erratic rainfall results in fluctuating production. This in turn leads to frequent scarcities, like the ones experienced in Indonesia and Vietnam in 1977 which created severe food shortages. Droughts in China in 1972, 1974 and 1985 brought depression of food grain production by up to about 25 million tones. Frequent droughts in India during 1966, 1968, 1972, 1974, 1979, 1982 and 1987 seriously affected the food and fodder production in the country. Hence, it is necessary to understand the distribution of South-West monsoon within the season to determine the extent to which the crop productions are likely to be affected by the vagaries of monsoon.

Several attempts have been made to understand the behaviour of South-West monsoon rainfall in different agro-climatic regions on the basis of historical rainfall records. These studies have brought out that (i) there is large variation in dates of commencement of South-West monsoon from year to year in different parts of the country, (ii) the monsoon rainfall is of sequential nature with long dry spells or breaks extends sometimes to the period of even one month or more, (iii) there is large year to year variation in dates of withdrawal of South-West monsoon, (iv) there is variation in quantum of rainfall received from year to year and (v) high intensity rainfall occurs in association with movement of cyclones or depression resulting in sizeable loss of rainwater through run-off and deep drainage.

Thus, crop production in dry lands fluctuates widely from year to year due to vagaries of weather. An aberrant weather can be categorized under three heads i.e. (i) delayed onset of monsoon, (ii) long gaps or breaks in rainfall, and (iii) early stoppage of rains towards the end of monsoon season. Therefore, to mitigate such weather situations, farmers should make some changes in normal cropping schedule for getting some production in place of total crop failure.

Crop Substitution

Alternate crop strategies have been worked out for important regions of the country for *vertisols, alfisols, entisols, submontane* and *sierozemic* soils. Strategy has also been evolved for normal onset of rains, breaks in rains, early withdrawal, its uneven distribution; through selection of uneven crops/varieties which are inefficient utilize of the soil moisture, less responsive to production input and potentially low producers should be substituted by more efficient ones. Appropriate crops, suiting varying rainfall situations, have been identified for most of the dry land regions of India.

Crops which do not under normal rainfall years may not do so under abnormal years. Studies conducted in agro climatic conditions of Varanasi (eastern U.P.) revealed that under normal monsoon crops like short duration upland rice, maize, pearl millet, blackgram, greengram, sesame, pigeon pea etc. should be taken up on the basis of needs. These crops should be followed by chickpea, lentil, barley, mustard, safflower, linseed etc. on residual moisture during winter season. If monsoon sets in as late as second week of July, short duration upland rice (variety-NDR-97 and NDR-118) may be included in place of Akashi and Cauvery is recommended.

If the rains are delayed beyond the period but start somewhere in last week of July or first week of August and growing season is reduced to 60-7-days, then cultivation of hybrid pearl millet (NHB 3-4, B.J. 104), blackgram (Type 9), greengram (var-Jagriti and Jyoti) may be included in pace of T-44 and k-851 etc. should be grown. Yet another alternative could be to harvest a fodder of either pearl millet, maize, sorghum or a mixture of cowpea, blackgram and one of the above fodder crops.

In case monsoon rains stop early towards the end of season, normal sowing of short duration upland rice, blackgarm and sesame may be taken up.

If the rain stops very early, i.e. by the end of August or first week of September, only fodder crops or grain legumes could be harvested. Depending upon the soil moisture condition, relay sowing of crops like chickpea, lentil, mustard, linseed and barley could be done in rabi season.

Table: *Traditional and Alternate Efficient crops in Different Dryland Regions of India*

S.N.	*Region*	*Traditional crop*	*Alternate efficient crop*
1.	Deccan Rabi season	Cotton, wheat	safflower
2.	Malwa Plateau	wheat	Safflower, Chick pea
3.	Uplands of Bihar Plateau and Orissa	Rice	Ragi, Black gram, Groundnut
4.	South-easy Rajasthan	Maize	Sorghum
5.	North Madhya Pradesh	Maize	Soybean
6.	Eastern UP	Kalitur	Chick pea
7.	Sierozems of Northwest India	Wheat	Mustard, Taramira (Eruca sativa)

During the recent drought, it was found that farmers in Karnataka, Andhra Pradesh and Maharashtra who went in for sunflower cultivation were in gainers. Sunflower succeeded where other crops failed. In other dry land regions, alternative efficient crops can profitably substitute the traditional ones.

Table: *Relative Yield of Traditional and Efficient Crops in Dryland Areas*

Region	*Traditional*	*Yield (q/ha)*	*Efficient crops*	*Yield (q/ha)*
Bellary	Cotton	2.0	Sorghum	26.7
Varanasi	Wheat	8.6	Chickpea	28.5
Ranchi	Upland Rice	28.8	Maize	33.6
Indore	Green gram, Wheat	11.8	Soybean	33.3
		11.0	Safflower	24.2
Agra	Wheat	10.3	Mustard	20.4
Hisar	Wheat	3.0	Taramira	16.0
Udaipur	Maize	18.0	Hybrid sorghum	29.0

Dry land research has remained confined to important traditional crops such as sorghum, millet, pulse and oilseeds and has not explored the possibility of growing non-traditional crops such as dye-providing crops {e.g. Henna (Lawsonia inermis: mehadi) and jaffra(Bixa ovellana) species (0e.g. cumin), and medicinal value crops (e.g. eitronella, lemon grass, senna and isabgol)}.

These crops need to find an important place in research aagenda of dry land farming.

Time has come for the relevant researchers to plan a joint integrated research programme for maximizing the profitability, productivity and sustainability of learning systems of rainfed areas. Sericulture offers great promise in rainfed farming strategy, particularly of the watershed approach in peninsular India.

Efficient Cropping System

Besides putting various measures to increase the productivity levels of dry land crops, efforts would also be needed to increase the cropping intensity in dry land areas which was generally 100%, implying that a single crop was taken during the year. Cropping intensities of these areas could be increased by practice of inter cropping and multi cropping (sequential) by way of more efficient utilization of resources. The cropping intensity would depend on the length of growing season which in turn depends on rainfall pattern and the soil moisture storage capacity of the soil. For example in Indore region, receiving 1000 mm annual rainfall, only single crop can be taken on shallow soils, inter cropping in medium depth soils and double cropping on deep soils. Similar crop combinations have been identified for different regions of the country. In dry land of Varanasi region upland rice-chickpea/lentil sequence can be practices with advantage.

Inter cropping of vegetables with grain crops was pursued vigorously in some centres such as Varanasi and Phulbani. At both the palces long duration pigeon pea was inter cropped with vehetables such as okra, radish and chilli. Such inter cropping systems would be very useful to get maximum returns from rainfall agriculture. Even at solapur, leafy vegetables and some short duration beans were grown as intercrops during the rainy season.

Fertilizer Use

Soils of dry lands in the country are not only thirsty but hungry also because these soils are severely eroded horizontly as well as vertically. Whenever efforts are made towards bunding and levelling of the fields in dry land areas, it is the surface soil which is removed. The resultant effect is that the fields are rendered shallow in depth and completely deprived of plant nutrients, particularly nitrogen, phosphorus and potassium. It is, therefore, necessary to apply all the three major nutrients in adequate amounts. Since soil moisture is limiting in dry lands, the availability of nutrients becomes limited, attempt should always be made to apply fertilizers in furrows below the seed. If seed-cum-fertilizer drills drawn by bullocks or tractors are available, this very objective can be fulfilled.

There has been belief among the farmers of dry land areas that use of fertilizer increases the chances of crop failure but recent findings have shown that the use of fertilizer is not only helpful in providing nutrients to crop but also helpful in efficient use of profile soil moisture. If dry land farmers are shown such results, they will be convinced to use ore and more fertilizers.

Studies on the management of legumes in crop sequences for their residual effect indicated that in alluvial soils an advantage of 25-30 kg N/ha could be obtained in barley or mustard grown after black gram or green gram. Another possibility for nitrogen management in cropping system is to use legumes as green manures either at flowering stage or after one picking. Studies conducted at Varanasi centres clearly showed that general yield levels of barley and mustard were greater when legumes raised in the previous season was incorporated I soil after first picking as compared to that harvested at normal maturity.

In dry land areas, a proper mixing of organic and inorganic would be desirable. Organics have low nutrient content, but help to improve the moisture holding capacity of soils. In addition to yield advantage, nutrients like potassium help to increase drought tolerance by affecting plant-soil relationship.

Table: *Nitrogen Economy to Legume-Cereal System (4 years average)*

Nitrogen level (kg/ha) Unicorporated	**Incorporated**	**Crop yield (q/ha)**
Green Gram	1.89	2.23
Barley		
0	16.98	13.65
30	21.30	18.64
60	24.43	21.84
90	27.27	25.20

Rain Water Management

Efficient management of rain water can boost agricultural production from dry lands. The broad bed and furrow system of the International Crop Research Institute for the Semi Arid Tropics

(ICRISAT) for managing rain water in vertisols made it possible to increase crop yields four to five times as compared to normal practice. However, this method could not be adopted widely by the farmers in India because it is costly and labour intensive. The vertical mulching developed at Bellary centres increases the infiltration of water in soil profiles and improves in situ moisture conservation. The scope for managing profile moisture is limited in alfisols but the surface run off in such soils can be reduced by ridge-and-furrow technique. Alternatively, application of compost and farm yard manure as well as raising legumes will add the organic matter to the soil and increase the water holding capacity.

The winter which is not retained by the soil flows out as surface runoff. The run-off-recycling holds immense prospects in deep black soils where the seepage losses are very much less. This runoff water, if not permitted to drain out safely, causes erosion. Therefore, safe disposal of excess water from the field drains to the disposal system should be planned properly. This excess runoff water can also be harvested in storing dug out ponds and recycled to donor area in the event of severe moisture stress during rainy season or for raising crops during the winter.

Water-shed Approach for Resource Improvement and Utilization

Watershed management is a holistic approach arrived at optimizing the use of lad, water and vegetation in an area and thus, providing solution to alleviate drought, moderate folds, prevent soil erosion, improve water availability and increase fuel, fodder and agricultural production on a sustained basis. On the basis of the experiences of ICAR Operational research Projects, which attracted the attention of our farmers, State departments, administrators and scientists, 47 model watersheds were established during the year 1983 for development, jointly by the Ministry of Agriculture, ICAR and various State Government Department and Agricultural Universities, in 16 states and then the Department of Agriculture and Cooperation launched the National Watershed Development Project for Rainfall Areas (NWDPRA) covering almost the same states. Out of these 47 model watersheds, the Central research Institute for dryland Agriculture (CRIDA). Hyderabad has been entrusted with 30 watersheds. These activities were in micro and mini-watersheds covering 500-2000 ha. Major components in these model watersheds are: (i) Improvement of water resources, (ii) In situ soil and water conservation: rain water harvesting for safe disposal of surface runoff,

(iii) increase in cropping intensity and (iv) alternate land use system for efficient use of lands as per land capability to provide stability in productivity.

The model watersheds in operation have provided a fruitful experience of how development can lead to all round improvement in food and fodder production, economic condition of the farmers. Sakho-majori model, where creation of eater source worked as a catalyst and triggered the development process can be repeated under similar situations. Similar experiences have been gained a Tejpura (Jhansi), Ariel (Bareilly District) and Tejpura watersheds which have been awarded the First and Second Prizes respectively by the President of India on 14-11-1988 based on the recommendation of National Productivity Council.

Alternate Land use System

All dry lands are not suitable for crop production. Some lands may be suitable for range/pasture management, while others for tree farming, ley farming, dry land horticulture, agro-forestry systems including alley cropping. All these systems which are alternatives to crop production are called as alternate land use systems. This system not only helps in generating much needed off-season employment in mono crop dry land but also minimizes risk, utilizes off season rains which may otherwise go waste as runoff, prevents degradation of soils and restores balance in the ecosystem.

Crop production may be disastrous in the years of drought, whereas drought resistant grasses and trees could be remunerative. Scientists of dry land have developed many alternate land use systems which may suit different agro ecological situations. These are alley cropping, agri-horticultural system and silvi-pastoral systems which utilize the resources in better way for increased and stabilized production from dry lands.

Alley Cropping: for imparting stability and providing sustainability to the farming system, a tree-cum-crop system will be one most appropriate for such situations. One such system called 'alley cropping'- a version of agro-forestry system, could meet the multiple requirements of food, fodder, fuel, fertilizer etc. Alley cropping is a system in which food crops are grown in alleys formed by hedge rows of trees or shrubs. The essential feature of the system is that hedge rows are cut back at planting and kept pruned during cropping to prevent shading and to reduce competition with food crops.

For example, fast growing leguminous trees such as *Leucaena leucocephala* or *liliricidia spp.* are planted in rows. During the cropping season, trees are lopped at about 0.5 metre height. These loppings are used as much to reduce moisture loss and improve the nutrient status of soil. Arable crops like maize, rice, pearl millet, legumes, oilseeds etc. are planted in the alleys formed by the two rows of threes. This is also known as agri-silvi culture. Alley cropping is also a form of conservation farming which enhances soil fertility and prevents erosion.

One very strong argument in favour of alley cropping is its ability to produce usable material even in years of severe drought. At Rajkot in 1985, rainfall received during the season was only 30% of the normal. There was total failure of grain production of the three legume crops tried in the system. In sole crop plots production was limited to 5.0 q/ha to 17.0 q/ha of green fodder. However, in alley cropped plots, *Leucaena* hedge-rows produced over 50.0 q/ha of green fodder. Agri-horticultural system: Agri-horticultural system palys an important role in dry land areas, especially in semi-arid regions where production of annual crops is not only low but also highly unstable. Fruit trees if suitably integrated in dryland farming system could add significantly to overall agricultural production including food, fuel and fodder, conservation of soil and water and stability in production and income. Dry land fruit trees being deep rooted and hardy, can better tolerated monsoonal aberrations than short duration seasonal crops. Hence, in drought year when annual crops usually fail or their production is highly depressed, fruit trees species yield considerable food, fodder and fuel.

A suitable example of agri-horti-system is growing of cow pea/ green gram/horse gram in inter space of *ber (Zizyphus mauritiaria)* at 6 x 6 m spacing at Hyderabad. Phalsa (*Grewia asiatica*) may be planted in between two ber plants in a row with a view to intensify the system. A well managed dry land orchard of *ber* should give 50 kg fruits per tree/year. There should be 250 plants/ha for optimized production. The grow income would touch around Rs. 50,000/ha (250 x 50 x 4), assuming that one kg *ber* fetches Rs. 4. One could get an additional income of Rs. 800-Rs. 1000 from green gram/cow pea (2.5-3.0 q/ha).

Silvi-Pastoral System: This system is suited to marginal dry lands and is most preferable where the fodder shortages are experienced frequently. The system essentially consists of a top feed tree species carrying grasses on legumes (preferable perennial) as understorey crops. Dry land farmers having larger holdings and keeping a land follow for a longer period for one reason on the other, should go in for this system which could provide both fodder and fuel. In a survey carried out in Andhra Pradesh, Karnataka and Maharashtra by CRIDA scientists, it was revealed that after food it is the fodder which is of paramount importance for sustaining animal wealth in rural areas. In years to come, fuel will assume greater importance.

In August, 1981 *Leucaena leucocephalla* was planted in contour trenches 7.5 m apart, the plant to plant spacing being maintained at 2.0 m at CRIDA. Four strips at upper reaches of plot (2% slope) were put under *Cenchrus ciliaris,* while lower four strips were seeded with *Stylosanthes hamata.* The system has come up very well.

Efficient Implements

In order to take full advantage of annual precipitation in dry land agriculture, higher doses of energy input is essential. Farmers in dry lands have been using traditional and outdated farm equipments which not only perform poorly but also demand a lot of energy and time and post-harvest operations. Farm implements can help to conserve as much rain water in situ as possible and to harvest rain water. Shallow off season tillage with pre-monsoon showers ensures better moisture conservation and lesser weed intensity. It has resulted in 20% yield increase in sorghum in Andhra Pradesh. Deep tillage helps in increasing water in soils having textural profiles and hard pan. This has resulted in 10% yield increase in sorghum and 9% yield increase in case of caster. For in-situ moisture conservation, land has to be opened so that it can cause hurdle to flow of rain water. Tillage machines of appropriate size and type matching the power sources need to be used. Location specific seeders have been developed for dry land areas and these have shown good prospects and promise. A feature of these machines is that the seeds and fertilizers are placed in the moist zone of the soil resulting in a high percentage of seed germination and good crop vigour. In deciding farm mechanization in dry land areas, where farmers are generally poor, and their socio-economic condition should always be kept in mind.

The foregoing discussions show that technology of crop production in dry land areas have been generated to a great extent. What is important now is to view it in socio-economic context of the farmers. Once the technology is adopted by the farmers, the contribution of dry land areas to the total production can be sizably improved and the living standards of the farmers of these areas can be improved. This has been clearly shown in selected watershed areas and what is needed is to have more watersheds identified, proper technology to be developed and implemented.

Unlocking the Hidden Potential of Dryland Agriculture

IF government statistics are to be believed, the Indian economy never had it so good. For the past several years, India has been one of the fastest growing economies in the world. The rate of growth of GDP recorded during the first three years of the tenth plan (7%) is the highest since planning began in 1951. Other macroeconomic indicators like the rate of saving, current account balances and inflation rates seem to indicate the economy is in good health.

Against this background, the story of Indian agriculture, in particular dryland agriculture, appears as an annoying twist in the plot. Available data show that 1990-2000 was not a happy decade for Indian agriculture. The overall growth rate of crop production during this decade was nearly half of what it was during the 1980s. The output of coarse cereals, pulses and oilseeds (covering about 45% of total cropped area and grown mostly in the drylands) fell during the 1990s and the rate of growth of their yields decelerated considerably.

All this re-emphasise the fact that the gap between irrigated and dryland agriculture has steadily widened, with the productivity of the latter being less than half of the former. This has been a direct and predictable consequence of the strategy adopted in the mid-1960s, whereby massive investment flowed to the already well-endowed regions and farmers of the country, leaving aside the poorer and less endowed regions.

Dryland agriculture emerges as the biggest drag on the growth of the economy. Indeed, the dominant strand of thinking among our policy-makers treats the drylands as a hopelessly lost bet. However, we should caution ourselves against such a hasty conclusion. Even at their low productivity levels, the quantitative significance of dryland agriculture is by no means small.

It accounts for 53% of total cropped area, 48% of the area under food crops and 68% under non-food crops. In terms of production, drylands account for nearly 80% of the output of coarse cereals, 50% of maize, 65% of chickpea and pigeonpea, 81% of groundnut and 88% of soyabean. Half the output of cotton in the country is from the dry districts (Shah et al., 1998). Given its large size and extremely low productivity levels, a unit rise in productivity in this sector is likely to have the largest impact on aggregate crop productivity. There is clear evidence that the yield potential of dryland varieties is much higher than what has been achieved on the farm. It is more appropriate to view the drylands as a source for future growth, a hidden potential waiting to be unlocked.

Realising the potential of drylands is by no means a simple task. Spread over nearly half of the country, the drylands cover cold arid regions, hot deserts, hilly and undulating uplands, forest areas, plateaus, ravines and coastal and non-coastal saline areas. They are the home to 43% of our population. Water availability, soil conditions and the length of the growing season show wide variations here. Nine states (Rajasthan, Madhya Pradesh, Maharashtra, Gujarat, Chhatisgarh, Jharkhand, Andhra Pradesh, Karnataka and Tamil Nadu) account for over 80% of the drylands. Annual rainfall in the drylands varies from less than 150 mm to 1600 mm. Soils vary from shallow skeletal soils of the deserts to medium to deep black soils.

The list of prominent dryland regions brings out their diversity:

1. Western Himalayas: Cold arid region with rainfall <150 mm and shallow skeletal soils;
2. Western Rajasthan, Kutch and northern part of Kathiawar peninsula: Hot arid region with rainfall <300 mm, desert and saline soils;
3. Rajasthan uplands (Aravallis) and Chambal districts of Madhya Pradesh: Semi-arid region with alluvium-derived soils and extensive land degradation leading to ravines;
4. Central Highlands, including Gujarat plains and western Madhya Pradesh (Malwa): Semi-arid region with rainfall of 500-1000 mm, medium and deep black soils;
5. Deccan Plateau, including Maharashtra and northern Karnataka: Semi-arid region with rainfall of 600-1000 mm, red and black soils;
6. Interior Andhra Pradesh (Telangana): Semi-arid region with 600 Â– 1000 mm rainfall, red and black soils;
7. Tamil Nadu Uplands and western Karnataka, semi-arid region with red loamy soils;
8. Sub-humid Eastern Plateau (Chhatisgarh), with rainfall of 1000 to 1600, red and yellow soils;
9. Sub-humid Eastern Chhotanagpur Plateau including Jharkhand, western Orissa and northern Andhra Pradesh, with rainfall of 1200 to 1600, red and lateritic soils.

The range and diversity of the drylands presents a qualitatively complex set of constraints which needs a more delicate mode of handling. We must give up the one-size-fits-all approach and focus on fine-tuning and matching our interventions to the subtle variations in local contexts.

At the outset, we must realize the crucial role the state has to play in a positive reshaping of the dryland landscape. In the current context of liberalization and dominance of anti-state sentiments (shared, paradoxically, by the staunchest of neo-liberals as well as the purest among radicals), this point needs strong emphasis.

In particular, the key role of kickstarting the growth process in the forsaken drylands has to be played by public investment. The drylands are caught in a low-level equilibrium trap. Public investment is required to enable them to break out of this trap. The endemic process of natural resource degradation needs to be checked. Public investment in drylands has to be substantial, multi-directional and sustained over a long period of time. What we see at the national level is a clear decline in public investment in agriculture since the mid-1980s. This trend needs correction and the state has to step up its investment in the drylands substantially.

The precise forms such investment should take vary widely from region to region. Rather than being imposed from above, they must organically develop through an active exchange with the local context and its problem typology. There are examples galore to show how imposed, inappropriate interventions cause havoc. The desert region of western Rajasthan is one of the hottest regions of the country receiving scanty rainfall (100 to 300 mm) every year. Soils are dry during most part of the year. There have been several major attempts to 'green' the desert landscape. The most celebrated of them is the Indira Gandhi Nahar Project (IGNP), a gigantic water transfer project of diverting water to the desert.

Today, excessive irrigation and poor drainage planning has led to extensive waterlogging and land degradation due to sodicity-alkalinity. There are reports of hard pan formation beneath the soil layer. The severe ecological damage of this project that entailed an attempt to extend the green revolution to the desert is a clear pointer to the limits of human action in mastering nature. Yet, mega-schemes like extending the IGNP canal or lifting and carrying water from Narmada are axes along which solutions to the water problem of this parched desert landscape are envisaged.

Interestingly, the desert has a long tradition of local water harvesting for drinking water purposes and limited amount of agricultural use. Public investment could strengthen these. *Tanka* (or underground cisterns located in the middle of a large circular area which is its artificial catchment), *nadi* (community ponds usually constructed in inter-dunal spaces where there is some surface water run-off because of underlying geological conditions), *dighi* (small square or semi-circular, step-reservoir) and *khadin* (embankment built across slopes in agricultural fields) are some of the traditional water storage structures of the desert.

Similarly, there are groundwater harvest structures as well, such as *baoris, kund, kuiya* etc. The desert has developed some locally appropriate agro-forestry systems and traditional systems of addressing the problem of wind erosion and shifting sand dunes. A clear understanding of the scope and limits of such systems is necessary before we impose our external solutions through ill-conceived mega schemes.

Many other parts of the drylands also have traditional water management systems, such as tank irrigation in semi-arid South India and southeastern Rajasthan and the *ahar-pyne* system of South Bihar plains. Without romanticizing 'tradition', we can say that these systems show minute awareness of variations in topography, soils, rainfall and crop needs. However, these have been designed for much smaller populations and their institutional support systems have decayed over time. Rejuvenating and strengthening these must be an important plank of policy intervention.

The experience of borewell irrigation technology in hard rock regions of western India shows that there can be no miracle solutions to the dryland challenge. The Malwa plateau in western Madhya Pradesh is a case in point. The semi-arid Malwa plateau has predominantly hard rock geology, being underlain by Deccan Trap basalts. The natural rate of recharge of groundwater in hard rock regions is very low. Hard rock aquifers can often be fairly large reservoirs of stored water (accumulated over several thousand years). However, once they dry up, they cannot easily be replenished. Increasing the depth of groundwater extraction creates a very real danger of groundwater mining. This means that great caution needs to be exercised in the extraction of groundwater here. Such a careful approach never seems to have informed the groundwater economy of Malwa.

Since the introduction of the green revolution package, aided by bore-well technology since the 1980s, farmers in the Malwa region have been on a hot chase of extracting as much groundwater as they can. Subsidised electricity has accelerated the process. At present, uncontrolled extraction has given rise to a severe, man-made crisis of groundwater here. 60% blocks in the region are classified as dark or overexploited, meaning that their levels of groundwater extraction are unsustainable. Falling water levels are reported from everywhere and the area is in the throes of a severe drinking water shortage during summer. Yet, groundwater mining through deeper drilling continues unabated.

The priority intervention here is replenishment of highly depleted groundwater aquifers through strategically located surface water storage structures and sub-surface dams. These supply-side interventions should go together with regulation of end-uses of water through community management of groundwater, low cost drip irrigation and diversification of the crop production systems using drought-resistant varieties and crops that require less water.

A different package of interventions needs to be adopted for medium-rainfall eastern India (covering Jharkhand, non-coastal Orissa and Chhattisgarh), where the problem is severe underutilisation of available water. These economies are characterised by high subsistence orientation, small size of landholdings, poor irrigation infrastructure and heavy outmigration. Within this region, the tribal pockets are more deprived than others. Varying land slopes, water availability and soil depth define several land situations here. Public investment should support a basket of community-based strategies focused around utilisation of available water resources to improve productivity in each land situation. Detailed grassroots work has shown this possibility. With plateauing of yields in the traditional green revolution belt, the rainfed agriculture of the eastern region holds the key to India's food security in the near future.

Yield is never the only objective in dryland agriculture. Dryfarming is an exercise in constrained optimisation of multiple objectives. These include household food security, fodder and firewood needs, minimum cash flow, use of available household labour etc. The risk of rainfall failure is the most enduring concern. Traditional cropping systems use diverse strategies like mixed-cropping and intercropping for rainfall insurance. The crop combinations and sequences are often highly complex and have come up taking into consideration minute variations in soil type, depth, crop maturity and susceptibility to rainfall fluctuations and household needs. As much as 60 crop mixes are reported to have existed in villages of peninsular India (Jodha, 1977). The number of crops mixed in one plot ranged from two to eight or even more. It is important to understand their value in terms of their contribution to stability of household income and maintenance of soil nutrient status. They are in great danger of being trampled over by mono-cropping practices once water becomes available.

We need not only viable agriculture packages but meticulously worked out land-use planning systems, which make careful use of available soil moisture through appropriate tree-crop mixes. Here, again, what works and what would fail will be known only after more detailed evaluation of such systems in the field. There are several dedicated institutes deputed to do this kind of work but their field presence is next to nil.

Another crucial area of neglect is livestock. Small and marginal farmers and landless labourers constitute almost two-thirds of these livestock-keeping households in India. Ownership of livestock has a crucial drought-cushioning role for small and marginal farmers. With increasing grazing regulations and encroachments of common land, availability of fodder from agricultural land is one of the central concerns of poor households, particularly in drought years. Hence, their seemingly 'irrational' preference for crops with a low grain/fodder ratio. In spite of the vital role of livestock in a rural household, strategies of improving livestock productivity and health have not been systematically integrated as central interventions in the drylands. Of particular importance is the neglect of small ruminants (goats), which are the major part of livestock holdings of poor households in drylands.

The range of required interventions extends well beyond dryland production systems. They encompass many other related areas and simultaneous action is required at all levels. Provision of cheap, timely and easily negotiable credit is one such area. Drylands are poor in terms of their banking networks. Informal credit markets with interest rates ranging from 60 to 120% dominate. Organisation into self-help groups (SHGs) and linking them with banks is an effective means of credit delivery to poor house-holds. Crop insurance can be one of the financial products of SHGs.

On account of their low marketed surplus, many dryland areas have poorly developed commodity markets. Distress sale is the order of the day here, often to the same trader who is also the provider of timely credit. Creation of marketing networks through checking distress sales and pooling of whatever surplus the area produces, could be an income generating activity, which a federation of SHGs could take up. A crucial problem here is that of creating additional storage facilities for agricultural produce at the local level. Banks could finance such an activity, as shown by their recent experiments with commodity futures through a system of warehouse receipts. A much larger issue, needing greater consensus at the national level, is that of effective implementation of a support price policy and local procurement by state agencies of dryland crops.

To sum up, addressing the challenge of dryland agriculture involves implementation of a package of several interlinked components.

* location-specific public investments in water infrastructure;
* soil enrichment and control of land degradation;
* agricultural package of locally appropriate seeds and low-cost, sustainable agricultural practices;
* strengthening livelihood options based on livestock, fisheries, agro-processing and forests;
* better support systems through credit, marketing, research and extension;
* mobilization of communities around natural resource rights;
* learning from local contexts about possibilities and limitations of different interventions. This applies to all actors, scientists or government representatives, NGO activists or a member of local communit

Facing the challenge of the drylands is no longer a matter of choice. It is an imperative if we are to meet the goal of national food security in the coming years. Even in the most optimistic scenario of further irrigation development in India, nearly 40% of national demand for food in 2020 will have to be met through increasing the productivity of rainfed dryland agriculture (Samaj Pragati Sahayog, 2006).

To maintain food security even at the current nutritional levels, 102 MT of foodgrains have to be produced additionally by 2020. Cropped area has plateaued in India since 1970. It has remained static at around the 140 million-hectare mark for the last three decades. This is no longer a source of increased output in Indian agriculture. As for irrigated agriculture, its contribution can arise from two sources: (*a*) expansion in the area under irrigation; and (*b*) yield improvements in the areas already under irrigation.

Table: *Projected Demand and Supply of Foodgrains in India in the Year 2020*

	(million tonnes)
Projected Food Demand in 2020	307
Average Food Production in Triennium Ending 2002	205
Gap to be met	102
Maximum Possible Contribution of Irrigated Agriculture of which	64
From Irrigated Area Expansion	38
From Increases in Productivity of Irrigated Agriculture	26
Minimum Balance required from the Rainfed Agriculture	38
Share of Rainfed Agriculture	37%

Source: Samaj Pragati Sahayog, 2006.

The ultimate irrigation potential has been estimated at 139 million hectares, of which 75 million hectares is from surface water and 64 million hectares from groundwater (Planning Commission, 2002). As per Land Use Statistics, India's gross irrigated area (GIA) was 75 million hectares in 2001, which left a balance of 64 million hectares yet to be exploited. Many states in northwest India have already exhausted their irrigation potential. Nearly 65% of this unutilised irrigation potential is in the eastern parts of the country, comprising the medium to high rainfall regions of West Bengal, Bihar, Jharkhand, Orissa, Chhattisgarh, eastern Uttar Pradesh and northern Andhra Pradesh. The GIA can be expected to reach a maximum of 100 million hectares by 2020. This is an increase of 1.25 million hectares per annum, comparable to what has been achieved historically between 1970 and 2000 (1.28 million hectares).

If irrigated area grows at this rate between 2000 and 2020, we would have an additional 25 million hectares under irrigation by 2020. Assuming the share of foodgrains in GIA does not fall, only an additional 16 million hectares of foodgrain would thereby come under irrigation by 2020. Now since much of this addition to irrigated area would be in eastern India, and given current low yields under irrigated conditions, even under a most hopeful scenario, irrigated yields here are unlikely to cross three tonnes per hectare by 2020. We can, therefore, expect expansion in irrigated area to contribute an additional 38 MT to the total annual output of foodgrains by the year 2020.

Another part of the additional food output could come from yield improvements in the areas already under irrigation. Out of the 75 million hectares of GIA, 46 million hectares were under irrigated food crops in 2000. Yields of irrigated agriculture in India began to plateau in the 1990s and have declined in some areas. Even if we optimistically assume that yield growth of 30 kg/ha is sustained over the next 20 years, the rise in yield by 2020 will be only 0.6 tonnes per hectare. Thus, areas already under irrigation could contribute an additional 26 MT of foodgrains to the shortfall in 2020.

The total contribution of irrigated agriculture to foodgrain production from both area expansion and yield improvements put together is, therefore, likely to be around 64 MT, still leaving a shortfall of 38 MT of foodgrains in 2020. In other words, even in the best possible scenario of irrigation development, about 40% of the additional supply of foodgrains needed to match future rise in demand will have to come from the unirrigated segment of Indian agriculture, most of which is located in the dryland areas. Thus, from the point of view of maintaining food security, there is no alternative to raising productivity of the drylands.

6

Management Steps to Improve Soil Quality

A basic soil audit is the first and sometimes the only monitoring tool used to assess changes in the soil. Unfortunately, the standard soil test done to determine nutrient levels (P, K, Ca, Mg, etc.) provides no information on soil biology and physical properties. Yet, most of the farmer-recognized criteria listed on the first page of this publication for healthy soils include, or are created by, soil organisms and soil physical properties. A better appreciation of these biological and physical soil properties, and how they affect soil management and productivity, has resulted in the adoption of several new soil health assessment techniques.

The USDA Soil Quality Institute provides a Soil Quality Test Kit Guide developed by John Doran and associates at the Agricultural Research Service's office in Lincoln, Nebraska. The kit was designed for field use. Components necessary to build a kit include many items commonly available such as pop bottles, flat bladed knives, a garden trowel, and plastic wrap. Also necessary to do the tests is some equipment usually not locally available such as hypodermic needles, latex tubing, a soil thermometer, an electrical conductivity meter, filter paper, and an EC calibration standard. The kit allows the measurement of water infiltration, water holding capacity, bulk density, pH, soil nitrate, salt concentration, aggregate stability, earthworm numbers, and respiration.

A greatly simplified and quick soil quality assessment is also available at the Soil Quality Institute's web page by clicking on "Getting to Know your Soil," near the bottom of the homepage. This greatly simplified method involves digging a hole and making some observations. Here are a few of the procedures shown at this website: Dig a hole 4 to 6 inches below the last tillage depth and observe how hard the digging was. Inspect plant roots for lots of branching and fine root hairs or a balled up condition. A lack of fine root hairs indicates oxygen deprivation, while sideways growth indicates a hardpan. The process goes on to tell about earthworms, smelling the soil, and assessing the aggregation.

A cropland-monitoring guide has been published by the Centre for Holistic Management. The monitoring guide contains a set of soil health indicators that are measurable in the field. No fancy equipment is needed to make the assessments described in this monitoring guide. In fact, all the equipment is cheap and locally available on almost any farm. Simple measurements can help determine the health of croplands in terms of the effectiveness of the nutrient cycle, water cycle, and the diversity of some soil organisms. Some of the assessments you can make using this guide are living organisms, aggregation, water infiltration, ground cover, and earthworms. The monitoring guide is easy to read and understand, and comes with a field sheet to record observations. It is available for $12 from the Centre for Holistic Management.

Some quick ways to identify a healthy soil include feeling it and smelling it. Grab a handful and take a whiff. Does it have an earthy smell? Is it a loose, crumbly soil with some earthworms present? Dr. Ray Weil, soil scientist at the University of Maryland describes how he would make a quick evaluation of a soil's health in just 5 minutes.

Look at the surface and see if it is crusted, which tells something about tillage practices used, organic matter, and structure. Pushing a soil probe down to 12 inches, lift out some soil and feel its texture. If a plow pan were present it would have been felt with the probe. Turn over a shovelful of soil to look for earthworms and smell for actinomycetes, which are microorganisms that help compost and stabilize decaying organic matter. Their activity leaves a fresh earthy smell in the soil.

Two more easy observations are to count the number of soil organisms in a square foot of surface crop residue and to pour a pint of water on the soil and record the time it takes to sink in. Comparisons can be made using these simple observations along with Ray Weil's evaluation above to determine how farm practices affect soil quality.

Simple Erosion Test

This test demonstrates the value of ground cover. Tape a white piece of paper near the end of a 3-foot-long stick. Hold the stick in one hand so as to have the paper end within 1 inch of a bare soil surface. Now pour a pint of water onto the bare soil within 2-3 inches of the white paper and observe the soil accumulation on the white paper. Tape another piece of white paper to the stick and repeat the operation, this time over soil with 100% ground cover, and observe the accumulation of soil on the paper. Compare the two pieces of paper. This simple test shows how effective ground cover can be at preventing soil particles from detaching from the soil surface.

Detachment of soil particles occurs when falling rainwater collides with bare ground. After enough water builds up on the soil surface, following detachment, overland water flow transports suspended soil down slope. Suspended soil in the runoff water abrades and detaches additional soil particles as the water travels overland. Preventing detachment is the most effective point of erosion control because it keeps the soil in place. Other erosion control practices which seek to slow soil particle transport and cause soil to be deposited before it reaches the stream are less effective at preventing erosion. These latter practices are the ones typically implemented (terraces and diversions). Terraces, diversions and many other erosion "control" practices are basically unnecessary if the ground stays covered year round. For erosion prevention, a high percentage of ground cover is a good "early warning" indicator of success, while bare ground indicates a high risk of erosion (Croplands monitoring guide, 98). Muddy runoff water and gullies are "too-late" indicators. The soil has already eroded by the time it shows up as muddy water and it's too late to save soil already suspended in the water.

Tools and Techniques to Build Soil

Can a cover crop be worked into your rotation? How about a high-residue crop or perennial sod? Are there economical sources of organic materials or manure in your area? Are there ways to reduce tillage and nitrogen fertilizer? Where feasible, bulky organic amendments may be added to supply both organic matter and plant nutrients. It is particularly useful to account for nutrients where organic fertilizers and amendments are utilized. Start with a soil test and a nutrient analysis of the material you are applying. Knowing the amount of nutrients needed to supply the crop to be grown guides the amount of amendment applied and can lead to significant reductions in fertilizer purchase. The nutrient composition of organic materials can be variable, which is all the more reason to determine the amount you have with appropriate testing. In addition to containing the major plant nutrients, organic fertilizers can supply many essential micronutrients. Proper calibration of the spreading equipment is also important to ensure accurate application rates.

Animal Manure

Manure is an excellent soil amendment, providing both organic matter and nutrients. Typical rates for dairy manure would be 10 to 30 tons per acre or 4,000 to 11,000 gallons of liquid for corn. At these rates the crop would get between 50 and 150 pounds of available nitrogen per acre. Additionally, lots of carbon would be added to the soil, resulting in no loss of soil organic matter. High crop residues grown from this manure application would also contribute organic matter.

However, a common problem with using manure as a crop nutrient source is that application rates are usually based on the nitrogen needs of the crop. Because some manures often have about as much phosphorus as they do nitrogen, this often leads to buildup of soil phosphorus. A classic example is chicken litter applied to crops that require high nitrogen levels, such as pasture grasses and corn. Broiler litter, for example, contains approximately 50 pounds of nitrogen and phosphorus and about 40 pounds of potassium per ton. A common fertilizer application for established fescue pasture would be about 50 pounds of nitrogen and 30 – 40 pounds of phosphorus per acre. If a ton of poultry litter were applied to supply the nitrogen needs of the fescue, an over-application of phosphorus would result. Several years of litter application can build soil phosphorus up to excessive levels. One easy answer is to this dilemma is to adjust the manure rate to meet the phosphorus needs of the crop and to supply the additional nitrogen with fertilizer or a legume cover crop.

Compost

Composting farm manure and other organic materials is an excellent way to stabilize their nutrient content. A significant portion of raw-manure nutrients are in unstable, soluble forms. Such unstable forms are more likely to run off if surface applied, or to leach if tilled into the soil. Therefore compost is not a good source of readily available plant nutrients like manures are. Compost releases its nutrients slowly, thereby minimizing losses. Quality compost contains more humus than its raw components because primary decomposition has occurred during the composting process. It also does not contribute the sticky gums and waxes that aggregate soil particles together as much as does raw manure because these substances are also released during the primary decomposition phase. Unlike manure, compost can be used at almost any rate without burning plants. In fact some greenhouse potting mixes contain 20 to 30% compost. Compost (like manure) should be analyzed by a laboratory to determine the nutrient value of a particular batch and insure its wise use. Composting also reduces the bulk of raw organic materials– especially manures which often have a high moisture content. However, while less bulky and easier to handle, composts can be expensive to buy. On-farm composting cuts costs dramatically compared with buying compost.

Cover Crops and Green Manures

Many types of plants can be grown as cover crops. Some of the more common ones include: rye, buckwheat, hairy vetch, crimson clover, subterranean clover, red clover, sweet clover, cowpeas, millet, and forage sorghums. Each of these plants has advantages over the others and their area of adaptability. Cover crops can maintain or increase soil organic matter if they are allowed to grow long enough to produce high herbage. All too often, people get in a hurry and take out a good cover crop just a week or two before it has reached its full potential. Hairy vetch or crimson clover can yield up to 2.5 tons per acre if allowed to go to 25% bloom stage. A mixture of rye and hairy vetch can produce even more.

In addition to the organic matter benefits, legume cover crops provide considerable nitrogen for crops that follow them. Consequently, the nitrogen rate can be reduced following a productive legume cover crop taken out at the correct time. For example, corn grown following 2 tons of hairy vetch should produce high yields of grain with only 50% of the normal nitrogen rate.

When small grains such as rye are used as cover crops and allowed to reach the flowering stage, additional nitrogen may be required to help offset the nitrogen tie-up caused from the high carbon addition of the rye residue. The same would be true of any high carbon amendment such as sawdust or wheat straw. Cover crops also suppress weeds, help break pest cycles, and through their pollen and nectar provide food sources for beneficial insects and honeybees. They can also cycle other soil nutrients making them available to subsequent crops as the green manure decomposes.

Humates

Humates and humic acid derivatives are a diverse family of products, generally obtained from various forms of oxidized coal. Coal-derived humus is essentially the same as humus extracts from soil but there has been a reluctance in some circles to accept it as a worthwhile soil additive. In part, this stems from a belief that only humus derived from recently decayed organic matter is beneficial. It is also true that the production and recycling of organic matter in the soil cannot be replaced by coal-derived humus. However, while sugars, gums, waxes and similar materials derived from fresh organic-matter decay play a vital role in both soil microbiology and structure, they are not humus. Only a small portion of the organic matter added to the soil will ever be converted to humus. Most will return to the atmosphere as carbon dioxide as it decays.

Many studies have shown positive effects of humates, while other studies have shown no such effects. Generally, the consensus is that they work well in low organic matter soils. In low amounts they do not produce positive results on soils high in organic matter. At high rates they may tie up soil nutrients.

There are many humus products on the market. They are not all the same. Humate products, should be evaluated in a small test plot for cost effectiveness before using. Sales people sometimes make exaggerated claims for their products.

Reduce Tillage

While tillage has become common to many production systems, its effects on the soil can be counter-productive. Tillage smoothes the soil surface and reduces natural soil aggregation and earthworm channels. Porosity and water infiltration are decreased following most tillage operations. Plow pans may develop in many situations. Tilled soils have much higher erosion rates than soils left covered with crop residue.

Due to all the problems associated with conventional tillage operations, acreage under reduced tillage systems is increasing on the American landscape. Any tillage system that leaves in excess of 30% surface residue is considered a "conservation tillage" system by USDA. Conservation tillage includes no-till, zero till, ridge-till, zone till, and some variations of chisel plowing and disking. These conservation till strategies and techniques allow for establishing crops into the previous crop's residues, which are purposely left on the soil surface. The principal benefits of conservation tillage are the reduction of soil erosion and improved water retention in the soil, resulting in more drought resistance.

Additional benefits, which many conservation tillage systems provide, include reduced fuel consumption, flexibility in planting and harvesting, reduced labour requirements, and improved soil tilth. Two of the most common conservation tillage systems are ridge tillage and no-till. Ridge tillage is a form of conservation tillage that uses specialized planters and cultivators to maintain permanent ridges on which row crops are grown. After harvest, crop residue is left until planting time. To plant the next crop, the planter places the seed in the top of the ridge after pushing residue out of the way and slicing off the surface of the ridge top. Ridges are re-formed during the last cultivation of the crop. Often, a band of herbicide is applied to the ridge top during planting. With banded herbicide applications, two cultivations are generally used: one to loosen the soil and another to create the ridge later in the season. No cultivation may be necessary if the herbicide is applied broadcast rather than banded. Because ridge tillage relies on cultivation to control weeds and reform ridges, this system allows farmers to further reduce their dependence on herbicides, compared with either conventional till or strict no-till systems.

Maintenance of the ridges is key to successful ridge tillage systems. The equipment must accurately reshape the ridge, clean away crop residue, plant in the ridge centre, and leave a viable seedbed. Not only does the ridge-tillage cultivator remove weeds, it also builds up the ridge. Harvesting in ridged fields may require tall, narrow dual wheels to be fitted to the combine. This modification permits the combine to straddle several rows, leaving the ridges undisturbed. Similarly, grain trucks and wagons cannot be driven randomly through the field. Maintenance of the ridge becomes a consideration for each process.

Conventional no-till methods have been criticized for a heavy reliance on chemical herbicides for weed control. Additionally, no-till farming requires careful management and expensive machinery for some applications. In many cases, the spring temperature of untilled soil is lower than that of tilled soil. This lower temperature can slow germination of early-planted corn or delay planting dates.

Also, increased insect and rodent pest problems have been reported. On the positive side, no-till methods offer excellent soil erosion prevention and decreased trips across the field. On well-drained soils that warm adequately in the spring, no-till has provided the same or better yields as conventional till. A recent equipment introduction into the no-till arena is the so-called "no-till cultivator." These cultivators permit cultivation in heavy residue and provide a non-chemical option to post-emergent herbicide applications. Farmers have the option to band herbicide in the row and use the no-till cultivator to clean the middles as a way to reduce herbicide use.

Minimize Synthetic Nitrogen Fertilizer Use

If at all possible, add carbon with nitrogen sources. Animal manure is a good way to add both carbon and nitrogen. When nitrogen fertilizer is used, try to do it at a time when a heavy crop residue is going onto the soil, too. For example, a rotation of corn, beans, and wheat would do well with nitrogen added after the corn residue was rolled down or lightly tilled in.

Spring planted soybeans would require no nitrogen. A small amount of nitrogen could be applied in the fall for the wheat. Following the wheat crop, a legume winter-annual cover crop could be planted. In the spring, when the cover crop is taken out, nitrogen rates for the corn would be reduced to account for the nitrogen in the legume.

The addition of legume residue would also be adding carbon. Avoid continual hay crops accompanied by high nitrogen fertilization. The continual removal of hay accompanied by high nitrogen speeds the decomposition of soil organic matter. Heavy fertilization of silage crops, where all the crop residue is removed (especially when accompanied by tillage), speeds soil decline and organic matter depletion.

Continue to Monitor for Indicators of Success or Failure

As you experiment with new practices and amendments, continue to monitor the soil for changes using some of the tools discussed in the Assessing Soil Health and Biological Activity section. Several of these monitoring guides have data sheets you can use in the field to record data and use for future comparison after changes are made to the farming practices. Review the principles of sustainable soil management and find ways to apply them in your operation. If the thought of pulling everything together seems overwhelming, start with only one or two new practices and build on them.

Crop Production Management

Pre-broadcasting Treatment for Better Groundnut Harvest

Groundnut is usually planted by broadcasting. Before broadcasting the seeds they are treated in three steps. First the groundnut is dried in the sun and graded. Drying seeds before peeling off can improve seed vigour. Usually the skin of the peanut is peeled off 4 days before broadcasting and then large size seeds are selected for broadcasting.

Then the seeds are soaked in cold water for 12-14 hours followed by again soaking in lukewarm water (35° C) for 2 minutes. The soaked seeds are then put in a bucket and a wet grass film is used to wrap them closely. The bud will sprout after 20 hours.

The third step is dressing the seed by pesticide. This is done by evenly splashing ammonium molybdate desolved in lukewarm water on the seeds and then dressing the seeds with duojunling (a local pesticide) in order to prevent disease and pests. For 50 kg of seeds 100 g ammonium molybdate and 250 g of Duojunling with a 25% concentration is required.

Controlling Number of Tillers

Rice yield can be increased by 30-40% by controlling the number of tillers. In this method the water is drained from the paddy fields so that the soil is exposed to the sun and allowed to dry till fissures appear. The fields are then irrigated again and this procedure is repeated several times till the number of tillers is only 80% of normal tillers of a variety. If there is more rain in tropical areas and there are less sunny days, there is no need to irrigate again after draining water until two leaf stage. The paddy field should be kept dry and wet alternately. The method produces 20% more grain bearing ears compared to the rate without the use of this method.

Another method, which can increase rice yield by 30%, is by frequent irrigation to the depth of two leaves. After the two leaf stage the field should is kept wet and dry alternately. In fertile clay soils more sunshine is needed other wise in sandy and less fertile soils long exposure to sunshine is not needed.

Method of Preventing Cotton Plant with non-blooming Buds

A successful method to solve the problem of cotton plants bearing non-blooming buds is to spray on the leaves a 0.1 % boric acid and 0.2 % borax hot mix diluted in cold water. The rate of spray is 6,975 liters/ha. The spraying is done during seedling stage, bud stage and blooming stages, respectively.

Rice Blast Diagnosis

There are two methods for diagnosis of rice blast in advance. One method is to take 6 rice leaves with some spots on them from the field. Put them in a bowl with a piece of wet cotton. A bowl is closed with a plastic cover. Put the bowl in an oven at 24°C for 24 hours. Then take the leaves out of the bowl and check whether there is a gray mould on the back of the leaf. If there is, then it can be confirmed that the rice plant in the field is suffering from blast.

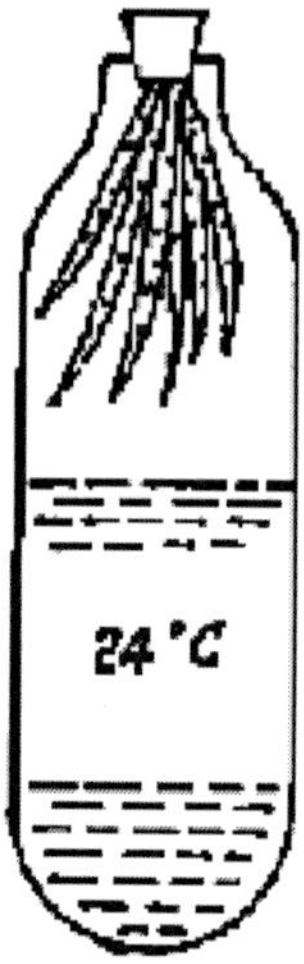

Another method is to take the 6 rice leaves with some spots from the field. Bind them and hang them above hot water at 40°C in a bottle. Then close the bottle for 24 hours. After that, check at the back of the leaves. The appearance of gray mould on it shows that the rice field will be attacked by blast. Preventive measures should be taken as early as possible.

Inter-cropping Corn and Chili for Increased Production

Inter-cropping corn and chili can enhance land productivity. The corn production not only remains the same but the chili crop also gets good profit.

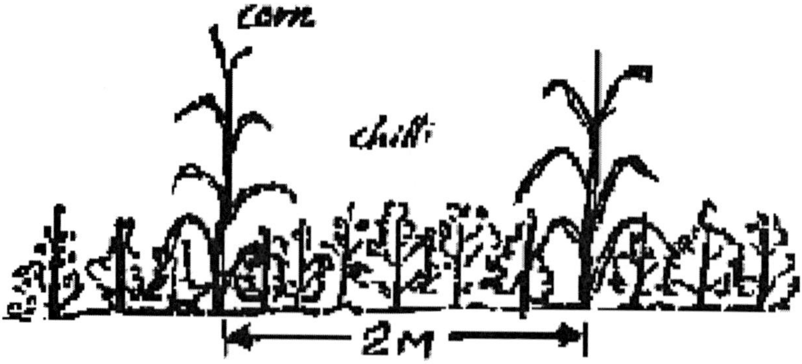

For land preparation, it is tilled to 25 cm depth and spread with organic manure. Now a days phosphorus and nitrogen (1:2/3) are also applied to the soil. The corn is planted at a row spacing of 100 cm × 15 cm with a plant density of 4500 rows per 0.15 ha. Two columns of chili are planted between corn in 30×30 cm spacing. The density of chili is at a rate of 4500 rows per 0.15 ha.

Now a day fertilizer is sprayed during crop growth. Weeding and pest control before corn cobs are formed require special attention. To promote chili to ripe old corn leaves should be removed during the growing stage of the corn. The heights of corn and chili plants are quite different. Lower leaves of corn shadow the chili, which is beneficial to it. On an average about 77 corn leaves are in 100-160 cm space, but chili occupies 60-80 cm space. This is a good arrangement to give sufficient light to chili. The corn production under this intercropping is same as of solo crop. Thus income from chili becomes extra. However, the method is suitable in summer season of semi-humid regions only as enough light and water are available at the same time.

Inter-cropping of Cotton and Melon (Cucumis Melo)

Compiled by Lui Xiaowen and Pong Shuping, Agriculture Extension Station, Zhibo, Sandong and Jia Fusheng and Sun Chuandian, Agriculture Extension Station, Bianhe, Sandong, China. For this inter-cropping arrangement, plant cotton in wide columns (80 cm) alternating with narrow columns (50-55 cm), with row spacing of 28-30 cm with a density is 3500 rows per 0.15 ha. Melon is planted in the space between wide columns of cotton with a row distance of 60-70 cm and a density of 800 rows per 0.15 ha for melon.

For good land preparation, in the winter season during plowing apply organic manure at a rate of 2500-3000 kg/ha, calcium phosphate 40-50 Kg/ha and urea 15-20 kg/ha. Then rake the soil for sowing in

April in the tropical regions of Sandong province of China. Pre-treat the cotton seeds in hot water at a temperature of 75°C for 2-3 days before sowing. The land is then covered with a plastic film after sowing. If the soil is dry it needs to be watered 1-2 liter/hole before sowing.

Select melon seeds of medium variety with high yield. Melon is sown at the same time with cotton. Put 2-3 melon seeds in each hole and cover with 2 cm of soil.

Management

Seeding stage: By controlling the cotton density, it grows faster and promotes the growth of melon. It needs about 80 days for melon to ripen from the time of sowing. During this period cotton grows slowly and does not need too much water and fertilizer. This creates a good growing condition for melon.

Thinning out of the cotton seedling is required at two leaf stage. Control of the cotton pests such as aphid, thrips and cutworm etc also. The soil should be loosened to keep the moisture intact and to promote growth. If melon grows too fast, only 3-4 branches and 3-4 melons should be kept. Its head and other branches should be cut. It is necessary to control pests such as powdery mildew.

Middle stage: Melon ripens in June-July. It should be harvested on time, since cotton needs more nutrition at this time. Chemical liquid should be sprayed twice to foster the growth of strong cotton seedling after it has 5-8 buds. The first spraying should be done at the end of June and the second spraying should be in the middle of July. Take out plastic film in the last ten days of June, and work the soil to loosen it in the spaces between the cotton columns.

Cotton blooming and ball formation stage: It is necessary to spray 10-15 kg urea before the 10th of July and around the 20th-25th of July. The head of the plant should be cut at the same time when the pruning is done during the first week of August. Attention should be paid to drainage, to loosening of the soil and to timely pest control.

Since both cotton and melon are drought-resistant and have different growing periods, inter-cropping the two can have high economic benefits. Natural resources such as land, light and water are used very economically. This technology is well suited to the rainfed mountain areas. However the intercropping is not suitable for cotton and melon varieties that have long growing period.

Honey bee-a good guard against bird damage to grape: Birds like to eat ripe grapes. Honeybee is a good guard against this damage. If beehives are put in the grape orchards during their grape maturity period, it very successfully protects the grape from bird damage. Bees attack birds, when they come to eat grape. According to local experience about 30 beehives/ha of orchard is enough for complete damage control.

Soil and Water Conservation: Intermittent Slope Level Ditch

Shaanxi Province is located in the loess plateau mountain area. Soil erosion is very serious here. With a long history of erosion control practices, the local people have found that the intermittent slope level ditch is very useful for erosion control and land utilization. It has all the advantages of intermittent bench terraces, silting dam (locally called Yudiba), and fish-scale pits. It helps to intercept sediment, store water and at the same time increases fertility.

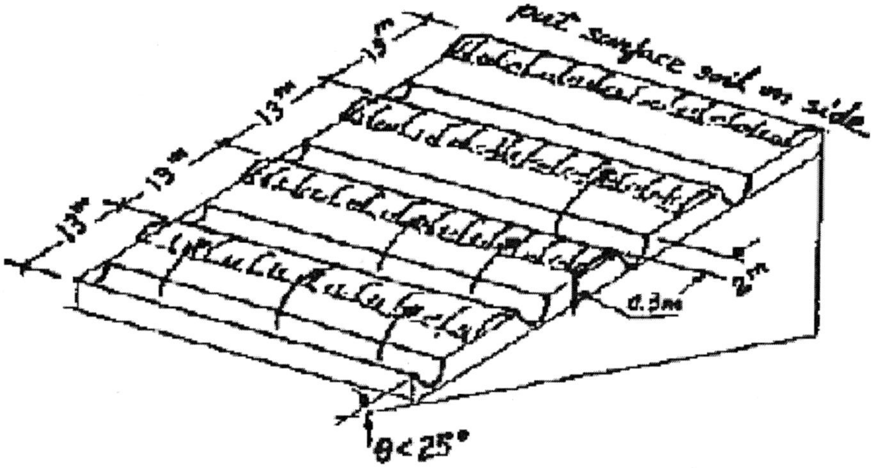

On slopes below 25°, the ditches are constructed 13 m apart along the contour,. The distance between the ditches is narrowed on slopes more than 25°. A ditch of 2 m width and 30 cm depth is first excavated and disposed on the inner side (upper slope); then on the outer side the earth is excavated to a depth of 1 m and the excavated material is disposed on the lower side for a ridge.

The ridge should be 0.7 m higher than the original slope surface with a bottom width of 1 m and top width of 0.5 m. The earth should then be rammed. Finally the excavated surface soil is refilled in to the ditch. The refilling should be 60 cm deep. A fast growing, economical tree should be planted where the earth has been refilled at a distance of 5 m from each other. Grasses and other cash crops can grow under the tree. In each hectare of slope land an estimated 750 m long ditches can be constructed.

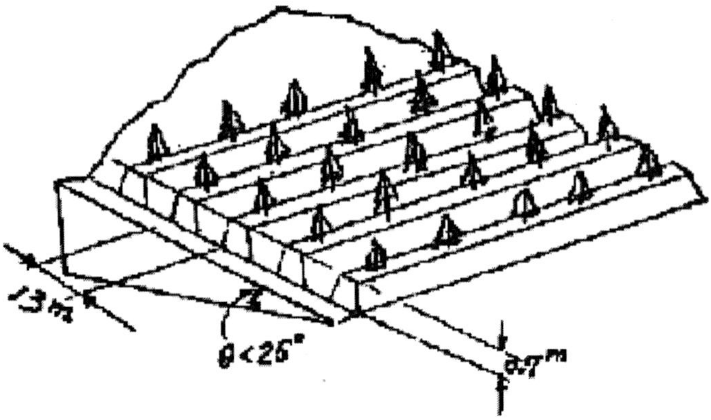

Advantages of the Intermittent Slope Level Ditch

Effectively controls soil and water loss: An area with 750 m/ha of constructed ditches can hold over 900 m^3 soil and water each time. Even a 100 mm rain can not cause any soil and water loss. The slope gradually turned into bench terraces with the natural scouring and yearly land preparation and maintenance.

Improve forest and grass growth: The refilled surface soil brings the fertility to normal top soil level, stores water and conserves soil. This raises the survival rate and growth rate of trees and grasses.

Multi-functions: The constructed ditches are not only favorable for the presently growing crops but also for the preparation of future transformation of slope land into forestland or grassland.

Low labour and other costs: On an average, each hectare of sloping land needs only less than 150 labour days and 75 Yuan (9 U.S. dollars) for tree seedling or grass seeds. Also the technique is very simple for the farmers.

The limitations of above ditches are that they needs lot of labour, consume much land and can reduce productivity in the first 2-3 years.

Southern China Farming Technology for Soil and Water Conservation

Some of the agronomic and cultural practices developed by the people in Southern China are given here.

Contour Cultivation on Terraces

Contour cultivation on terraces is practiced on a large scale for soil and water conservation. It has the capacity to retard runoff, increase infiltration of rainfall and conserve soil and water.

Ditch and Ridge Tillage

This is a technology used in mountainous areas of south China. People plough the sloping lands into contour ditches or into ridged farmland. This increases surface roughness to retard and store runoff, which benefits the growth of crops. The main patterns used in south China are given below.

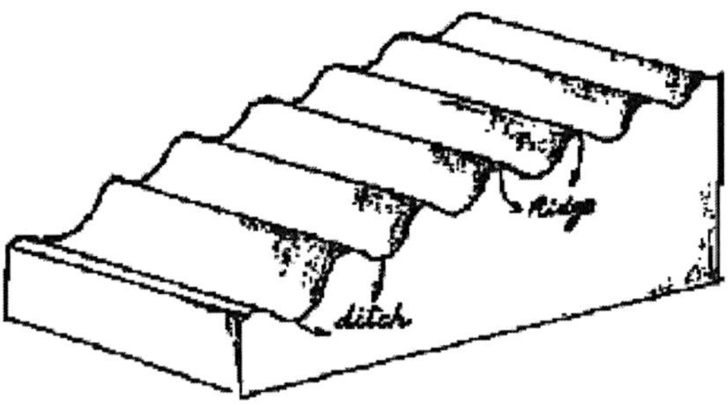

Ditch Tillage for Preserving Soil Fertility

The sloping land is excavated into ditches with width and depth along the contour as required for the row spacing of crops or trees. The top soil is put back with base fertilizer into the ditch.

Contour Ridge Tillage: In this type of tillage a ditch or ridge along sloping land contour is constructed to shorten the slope length and change the direction of runoff flow for the purpose of storing water, preventing scouring and combating drought and soil erosion.

Pit Planting

This technology is widely used in the steep and arid sloping land. Pits of needed size are excavated along contour. The excavated top soil is disposed on the upper side of the slope and kept for refilling. More sub-surface soil is excavated and disposed on the lower side of the slope to form a half-circle ridge for storing water and preventing soil erosion. Then, the stored top soil on the upper side is refilled in the excavated pits along with base fertilizers.

The excavated pits between upper and lower part are staggered in order to shorten slope runoff line, increase capacity for storing water and enough space for crop growth.

Semi-dry Rice Farming System

In this system the large paddy fields are changed into ridged and ditched fields to store water in the ditches and transplant rice seedling on the ridges. The advantage of this system is that soil erosion is reduced effectively and the yield of paddy rice increased.

Multiple Crop Rotation

Three crops in a year are grown in south China. It is also called the multiple crop rotation system. For instance, wheat-maize-sweet potato or wheat-yellow bean-sweet potato are grown. It has the advantage of conserving soil and water and increasing crop yield. It makes full use of climatic resources, increases the area for photosynthesis, and prolongs the duration of surface cover.

Minimum Tillage and Mulching System

The characteristic of this system is that the farmland is not ploughed after harvesting the previous crop. Ditches are excavated in rows with fixed spacing for broadcasting. The crop residues between ditches are retained and the previous crop straws or leaves or green manure are used as mulch to cover the newly planted crop. This system has the advantages of reducing soil erosion, reducing loss of organic matter and increasing crop yield.

Contour Strip Inter-cropping System

This system is suitable for sloping lands in south China especially with slopes below 25°.

Grassland strip rotation system: Grassland strips in rotation with crops can effectively control soil erosion, intercept runoff, increase infiltration of soil, improve water quality, and increase crop yield on the sloping lands. The nitrogen fixing grassland rotated with crops will improve soil fertility and structure.

Technique to Pave Gravel on the Sloping Lands for Soil and Water Conservation

Gravel-cover farming is a non-tillage technology used for a long time by people of Gansu Province. Gansu Province is located in the semi-arid and arid region with low rainfall and lack of water resources for irrigation. This is a typical rainfed farming system. In this technique a certain thickness of gravel is paved on the surface soil of the sloping land to store water, reduce evaporation, preserve fertility and reduce soil temperature, sanitation and soil loss.

By this technique, the requirements of water for crop growth can be basically met even under semi-arid and arid climatic conditions without irrigation. Since, the rainfall is generally concentrated in July, August and September, agriculture production is not guaranteed without irrigation. However, by paving gravel in a scattered manner on the surface of sloping lands, soil and water are preserved which can assure a harvest, increase crop yields, increase infiltration, reduce evaporation, temperature and salinization and prevent erosion. The example below illustrates the technique.

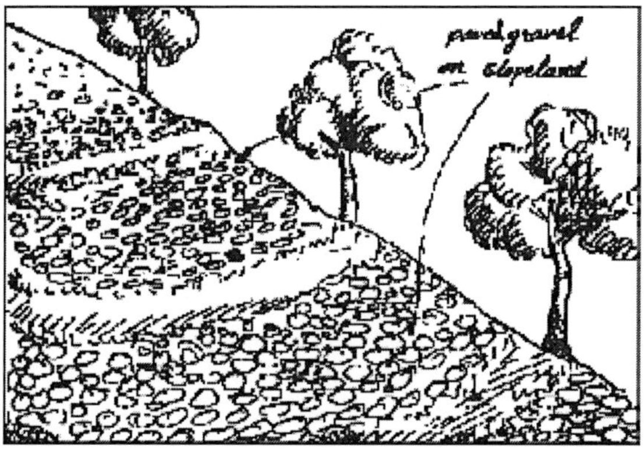

The sloping land is first ploughed once and the soil raked after the fall season. Then 75, 000 kg of organic matter for melon production as well as 150 kg of urea and 300 kg of ammonium calcium in each hectare are applied as base fertilizer. Pits are dug in rows along contour according to the row spacing of melon. The pits should be marked so that the gravel can be put around. Then between rows, some vegetables such as pepper or eggplant can be inter-cropped. Finally, a layer of gravel in 5 cm thickness can be paved on the remaining sloping land. The pits can be used for growing melon during the right season. After the harvest of melon, the second crop, for instance, maize or bean can be grown without tillage.

The gravel-covered farmland can be paved once and used for 15 to 40 years without tillage.

Method of Planting Trees on Terrace Risers

Tree species, which are suitable for planting by inserting them on terrace risers, are used. Strong 1-3 year old seedlings with thickness of 0.5-2.0 cm should be selected and cut into strips of 20-30 cm length. A 3 cm diameter, 60 cm long sharp wood is used to make hole on the terrace risers with a spacing of (0.2m-0.50m) × (0.5m-1.0m). The selected tree stripes are then inserted and the whole is filled back with wet soil. The terrace riser plantation stabilizes terraces and generates income from trees but is limited by rainfall conditions.

Amorpha Fruticosa-a Good Tree Species for Soil and Water Conservation

This species is found in reasons with 1,000 m AMSL. It is widely distributed in northeast China, north China, Guangdong, Yuannan and Guangxi Provinces. It has a well developed root system, is resistant to drought, and requires cold and wet conditions to grow and can survive on saline and infertile soils, seriously eroded gullies in both dry and humid climates. It is of high economic value as it is used for weaving, fodder, fuel wood, manure etc. It retards runoff and conserves soil.

A good Species of Herb for Soil and Water Conservation

Buffalogourd *(Cucurbita foetidissima)* is a kind of perennial herb of the gourd family with perennial roots. It can survive 30-40 years. The plant promises high oil bearing herbal fruits from the very first year to the end and tuber crop underground. It is therefore a kind of economic plant with multiple applications, high adaptability, readily cultivated on barren mountains and floodplains, which are otherwise not arable. An annual intensive farming and fallow cycle is used. Bringing the plant to mountainous areas in north China, which is extremely dry and sparsely vegetated, may be rewarding in conservation of soil and water, while at the same time increasing the income of local people by the sale of herb.

Method of Building Level Strips of Land by Contour Plowing

Sloping lands can be changed in level stripes by contour plowing in about 5 year period. The ridges formed by this method in between two strips can be further reinforced by planting bushes, fodder grass or medicinal plants on them. This adds to more economic benefits. For Inner Mongolia region to develop level strips in the sloping lands below 10° contour plow the lands. On the lower side, the height can increase by 20-24 cm and on the higher side the height can decrease by 20-24 cm. The slope becomes gradual after plowing several times. Experience shows that a 7 m wide belt needs to be contour ploughed 3-4 times. A 10 m level land needs to be ploughed 10 times.

Covering Crops with Coloured Plastic Film for Increasing Agriculture Production

Crops like rice seedling and carrot grow faster if covered by a red plastic film. Red colour strengthens the absorption capacity of the leaves and makes the seedling grow healthy which in turn improves the harvest. If a yellow plastic film is used to cover bean, lettuce, celery and cucumber, they will grow well and increase the yield by 50-100%. Farmers can get good harvest and prolong the marketing supply period to get economic benefit. If the temperature gradient of the green house is to be adjusted, silver coloured film should be used.

Good Way to Preserve Agro-plastic Film

There are 3 indigenous methods to preserve and save agro-plastic film

1. Put the clean plastic film in a plastic bag and bind the mouth of the bag and keep in a cool and damp place. This way the film can be preserved for 2 years.

2. Put the plastic film in a big jar with a weight on it, then add water to the jar until the film is submerged and cover it. This way the film can be preserved for 4 years.

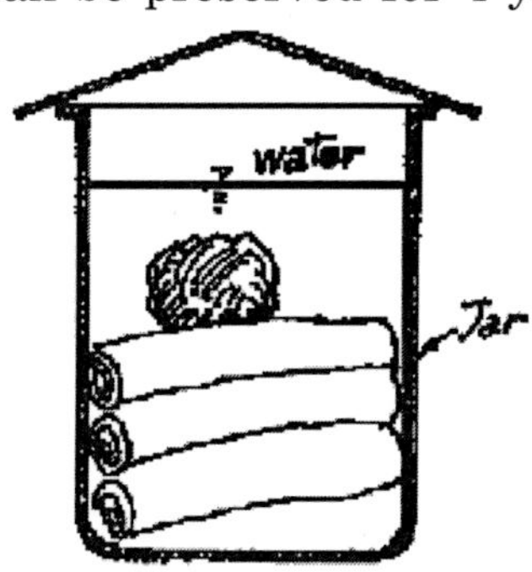

3. Put the clean plastic film in a plastic bag and store it in a 30 cm deep hollow or a cellar. This way the film can be preserved for 3 years.

7

Agro-climatic Zones and Crop Classification

Meteorological conditions before, during and after application of agrochemicals are very important for both efficacy and crop safety. Also, the speed of growth and development of plants and crops, and of pests, is strongly influenced by meteorological parameters. Although micrometeorological circumstances can differ within small areas and although micrometeorological data is recorded before, during and after application of products in efficacy trials, it is better, when comparing areas, to use data on climate, because: "Climate is the synthesis of weather events over the whole of a period statistically long enough to establish its statistical ensemble properties (mean value, variation, probabilities of extreme events, etc.) and is largely independent of any instantaneous events."

In order to assess the importance of climate at different localities for any kind of trials, it is necessary to establish what relationships exist between the growth of specific crops and their pests and the values of certain climatic elements. Climatic elements of high importance for the efficacy and phytotoxicity of plant protection products, for the germination and growth of weeds and for rate of development of other pests are: temperature (mean, minimum, maximum), relative humidity, precipitation, radiation and wind (direction and velocity).

Most crops show distinct differences, within their area of distribution, in many of their properties, for example in phenological data, growth, yield and quality. The chief cause of such differences is to be sought in variations in climate. Even fairly small variations in a single climatic element can exert an obvious influence, for example in relation to cultivars of a given plant.

These differences need only occur during a few growth controlling periods, and plants are affected not only by differences in the single climatic elements but also by various combinations occurring simultaneously or in succession. For this reason, the usual division of Europe into a few areas with distinct values of the main climatic elements is inadequate from the point of view of agronomy. By taking into account all the climatic elements and making a more detailed classification of their respective values, it is possible to divide Europe into numerous subareas, each with its narrowly differentiated 'general type of agroclimate'.

Variety of Climates in Europe

Europe extends over 30 degrees of latitude in a north–south direction, which means that incident sunshine releases very different quantities of energy in different areas of Europe, particular during certain seasons. The very broken coastline in the south, west and north also plays an important part in determining the distribution of temperature and rainfall from west to east. All climate maps shows this multiplicity. In relation to crops, the main points of interest are: rainfall (during the growing period), temperature (mostly only temperature in the growing period, but for some crops, temperature throughout the year), frost-free periods and humidity.

Rainfall

Average annual rainfall is abundant nearly everywhere in Europe and in some areas of the Middle East. However, in the mountains, it is much heavier than in the surrounding lowlands (Walter, 1969). Prevailing winds over most of the area are westerly all the year, but they are stronger and more constant in the northwest. The maritime precipitation furnished by these prevailing moist westerly winds is blocked by coastal mountains in several places, but this only affects the climate locally and does not act as a complete barrier to the precipitation.

The east–west European system of mountains from Spain to the Balkans is, however, an important climatic divide in that it acts as a barrier to the Mediterranean climate from the south, as well as to the maritime climate from the west and northwest. The broad pattern of central Europe is one of transition from the oceanic northwest of Europe to continental Russia. From west to east, precipitation falls gradually (up to 40% reduction), except where highlands and mountains intervene.

Temperature and Frost-free Period

Temperature in Europe varies greatly along any parallel of latitude, owing to prevailing winds and variation in altitude. Average temperature decreases as the altitude increases and the rate of decrease is about the same everywhere. It falls about 1 °C for every 100 m of elevation. In northern and northwestern Europe, and in most of central Europe, the prevailing winds are from the west, and blow straight from the warm ocean, which is a more important source of heat to western Europe in winter than the direct rays of the sun. These winds are the cause of the mild, moist and cloudy climate of western and northern Europe.

A notable characteristic of the European climate is its lack of extreme variations in a particular area. Along most of the west coast of the continent, the difference between the mean temperature of the warmest and coldest months is less than 6.5°C. During the winter months, the centre of Europe is cold, but the northwestern part has a relatively high temperature because of the warm air currents coming in from the ocean. The coldest part of Europe in January is in the northeastern part of European Russia and the North of Scandinavia, while the warmest areas in winter are the Mediterranean islands and the tips of Mediterranean peninsulas. In some northern regions of Europe and Asia, the amount of radiation is a limiting factor in the growing season. The term 'frost-free season' is generally used to designate the number of days without killing frost in any region. It is determined strictly by the prevailing temperature conditions, and only expresses what may be called the 'thermal growing season' of a large area. It does not take into account local environmental factors, such as lack of moisture or poor soil. These may prevent a specific crop from being grown in an area, even where the season is satisfactorily frost-free.

Climate Comparability Definition of Zones

Many authors have published climate classification systems and maps showing the limits of the various climates, main climatic types, climatic types, climatic subtypes and transition climates. A review of this work was given by Essenwanger (2001). Classification of climate is a grouping of atmospheric conditions for locations which show similar climatic conditions (climate types) separated by defined boundaries applied to one or more meteorological elements (Essenwanger, 2001).

None of these classification systems is based on all climatic elements affecting the crop. A map was accordingly constructed showing all boundaries between areas of different climates within Europe. In this way, it should be possible to use all climatic parameters influencing crops and pests. In mountainous areas, the variety of climatic factors makes it impossible to give a general characterization for the entire zone.

Climate Diagrams

Walter & Lieth (1960) produced climate diagrams from many meteorological stations throughout the world. These are brief summaries of average climate variables and seasonal variation.

The diagrams display monthly averages for temperature and precipitation and the seasonal patterns over a year. They also give an idea of the duration of cold/warm or wet/dry seasons and the extent of their fluctuations. Since all the diagrams are plotted on the same scale, it is possible to compare moisture, temperature and other environmental conditions in widely separated parts of the world.

The boundaries between the 'general agro-climate' subareas hardly ever represent an abrupt transition from one climatic element to another. Besides, these classification systems are not based on all the climatic elements which affect crops, and in most cases they make no contribution to the solution of agricultural problems. Hence the present division into 'general agro-climate' subareas does not represent any new grouping but is a 'projection' on a single map of the divisions as conceived by Thran & Broekhuizen (1965).

Climate Classification System

Köppen & Geiger (1936) made a climate classification system based on a quantitative system of temperature and amount of precipitation. The Köppen system distinguishes six major types of regional climate (A, B, C, D, E, H), which are further divided into subcategories. The subcategories BW and BS take vegetation into account, and a further subdistinction between 'k' and 'h' distinguishes hot (h) from cool (k) climates.

According to the Köppen system, the European and Mediterranean region (EPPO region) belongs to regional climates C (moist climates with mild winters), D (moist climates with severe winters) and H (highland climates). Plants were used as meteorological instruments for measuring precipitation effectiveness and drainage to make the regional climates : Bhw, BWh, Cfa, Cfb, Csa, Csb, Cfa, Dfa, Dfb, Dfc.

One of the modifications of the system of Köppen & Geiger was made by Strahler owing to air-mass source regions and their zones of contact. Strahler's modification is based on latitudinal and coastal locations of places. He developed three groups of latitude climates, divided into 15 groups in all. This is a modification of a plant distribution system that first appeared in the USDA Yearbook, conforming to the principle that the presence and form of natural vegetation are mainly attributable to the qualities of the atmosphere. An easy way to recognize a climatic region is through its effect on the predominant groupings of plants that grow on the surface of the earth. An important modification of the basic principles of Köppen has been made by Thran & Broekhuizen (1965), by placing emphasis on the relevant time of year when evaluating climatic data. Thus, climate data for the winter season is not relevant for the pest problems of spring-sown crops. Only when the winter period is of significant importance to weeds (overwintering, emergence), insect pests (diapause, mortality) or pathogens (overwintering bodies) in relation to the autumn-sown crop or to a long-term crop, is it of major importance to look at the climate data for the full year.

Thran & Broekhuizen (1965) tried to create new agroclimatic subareas, based on the length of the period from spring to midsummer, and from midsummer to autumn. These 'phenological' periods, which provide a suitable basis for evaluating agricultural potential, refer to the period between the sowing of the earliest crop in spring to the sowing of the latest crop in the autumn. It gives some information on the time when spring begins (e.g. when the +5 °C mean daily temperature is exceeded and conditions remain frost-free) and the length of the vegetation period. Using the Köppen & Geiger classification system, Thran & Broekhuizen adjusted annual precipitation, annual temperature, summer temperature (warmest month), winter temperature (coldest month), precipitation and photoperiod in up to nine ranks. With the help of this ranking system, they developed 76 different agro-climatic subareas in Europe.

Proposed Zones

On the basis of all the information described above, and taking into account natural vegetation maps for Europe and Asia, we have been able to divide Europe, the Mediterranean area and the Middle East into a number of agro-climatic zones 3.

The Mediterranean zone comprises the countries or parts of countries around the Mediterranean sea, together with Jordan, Macedonia and Portugal. It is the region of 'Mediterranean' plant species, favoured by mild winter and warm summer temperatures with relatively wet winters and dry summers. This corresponds to the region with codes Csa and Csb according to Köppen & Geiger, code IV according to Walter & Lieth, and codes 33–36, 458, 65–75, V–Y according to Thran & Broekhuizen.

The Maritime zone is the zone north of the line from the coastal zone of southwest France, through Lyon (France), to the south border of Switzerland and Austria, west of the border between Austria and Hungary, west of the border between Czech Republic and Slovakia, west of the river Oder (between Poland and Germany). This zone also includes Ireland, Sweden and the United Kingdom. It is the region of 'Atlantic' plant species, which grow in moderately cool or cold winters and fairly mild summer temperatures, with relatively wet winters and wet to occasionally dry summers. This corresponds to the region with codes Cfa and Cfb according to Köppen & Geiger, code VI according to Walter & Lieth, and codes 3, 6, 7, 8, 14b, 17, 19–25, 27–33, 39–45 according to Thran & Broekhuizen.

The *North-east zone* includes the countries and the regions east of the river Oder (between Poland and Germany), north of the border between Czech Republic and Poland, west of the border between Poland and Ukraine, north of the border between Ukraine and Belarus, Russia north of 50°latitude. It is the region of 'Continental' plant species that grow in cold relatively wet winters and mild dry summers. This corresponds to the region with codes Dfb and Dfc according to Köppen & Geiger, code VIII according to Walter & Lieth, and codes 1, 4, 5, 14, 18, 60, A, B, E, F, G, I, M, H according to Thran & Broekhuizen.

The *Central zone*is made up of Bosnia-Herzegovina, Bulgaria, Croatia, Hungary, Moldova, Romania, Russia south of 50° latitude, Slovakia, Slovenia, Serbia and Montenegro, Turkey, Ukraine, except the Mediterranean coastal zones. It is the region of 'Continental' plant species that grow in cold relatively dry winters and warm dry to occasionally wet summers. This corresponds to the region with codes Cfb, Dfb and Bsk according to Köppen & Geiger, code VII according to Walter & Lieth, and codes 60–64, P, R, S, N, K, L according to Thran & Broekhuizen. This zonation is proposed for the comparability of efficacy evaluation data. A similar analysis can be made for data on residues of plant protection products, and results in only two zones (Northern Europe and Mediterranean).

Other Relevant Factors

Climate is only one of the factors to be considered in establishing the relevance of data on the evaluation of plant protection products from one part of the EPPO region to another. Some of the other factors that may need to be considered are presented below. It is not possible to provide a prescriptive list, as this depends on the individual circumstances of use, mode of action, physical and chemical properties of the product, etc. Instead, the case for comparability should focus on factors relevant for performance and crop safety, and for the biology and incidence of the target. Extrapolation between regions may still be valid when the factors differ significantly, provided that the situation, in terms of efficacy and crop safety of the plant protection product, can be considered to be at least as challenging in the area of 'recognition' as in the area of registered use.

Edaphic Conditions

For soil-applied products, it is important to compare soil types, organic matter content and pH. It may also be important to take into account soil moisture content/deficit over the duration of the trial.

Agronomic Conditions

Cultural conditions and agronomy which may need to be considered include soil cultivation, application methods, cultivars, fertilizer regime, times of planting and harvest.

Differences in Target Pest Pressure

Certain aspects related to the target pests may give rise to a more or less harsh test of the efficacy and crop safety of the plant protection product. These could include: differences in epidemiology or population dynamics of the pests, different races of the target pest, resistance of pest populations.

Ecological Divisions of India-Agricultural Meteorology

A systematic appraisal of the soil and climatic resources is a pre-requisite for formulating effective land use plan for various regions of our country. Mapping of the various agro-ecological regions will help in identifying suitable cropping patterns for a particular region. There are several classifications of agro-climatic regions and soils proposed by several agencies. The reader is advised to study the earlier articless on the related topics Land Capability Classification and Agro-climatic Divisions of India. This article is on the agro-ecological regions of India; but there may be several things similar to the article on Agro-climatic Divisions of India. We must keep in mind the distinction between agroclimatic divisions and agro-ecological regions even though there may be similarities.

Indian sub-continent exhibits a variety of landscapes and climatic conditions which are noticeable in the types of soils and vegetation. Needless to say that there exists a significant relationship between climate, soils, type of land and vegetation. While preparing for land use plan or a cropping pattern or agricultural or animal husbandry system we should keep In mind the combine effect of climate, soil, land formation, topography and vegetation of the place.

In the past, several attempts have been made to understand and classify the agro-climatic regions as well as the agro-ecological regions of our country. In the year 1954, Carter divided India into six climatic regions, ranging from arid to perhumid, based on the criteria of Thornthwhite system of climatic classification. Murthy and Pandey (1978) on the basis of physiography, climate (rainfall and potential water surplus/deficit), soils and agricultural regions identified eight agroecological regions. Though this was a good beginning this system of classification suffered from some limitations. Based on the data of 160 meteorological stations in the country and using the concept of moisture adequacy index Subramaniam (1983) proposed 29 agro-ecological zones with the possible 36 combinations of moisture adequacy index and dominant soil groups following the pattern of FAO/UNESCO. Sehgal et al.(1987) prepared a computerized bio-climatic map of North West India, based on the criteria of dry month (the month having the real moisture deficit of 40% or more of the potential evapo-transpiration-PET). Krishnaft (1988) delineated 40 soil-climatic zones based on major soil types and moisture index.

Planning Commission as part of the mid-term appraisal of the planning targets of VII Plan (1985-1990) divided the country into 15 broad agro-climatic zones based on physiography and climate. They are the following: 1. Western Himalayan Region, 2. Eastern Himalayan I Region, 3. Lower Gangetic Plains Region, 4. Middle Gangetic Plains Region, 5. Upper Gangetic Plains Region, 6. Trans-Gangetic Plains Region, 7. Eastern Plateau and Hills Region, I 8. Central Plateau and Hills Region, 9. Western Plateau and Hills Region, 10. Southern Plateau and Hills Region, 11. East Coast Plains and Hills Region, 12. West Coast plains and Hills Region, 13. Gujarat Plains and Hills Region, 14. Western Dry Region and 15. The Island Region.

The state agricultural universities were entrusted with the task of dividing these agroclimatic zones into state-wise sub-agro-climatic zones under the National Agricultural Research Project (NARP). Accordingly 127 sub zones were identified based on the rainfall, existing cropping pattern and administrative units.

The government of India was not satisfied with the classifications done so far. It further entrusted the National Bureau of Soil Survey and Land Use Planning (NBSS&LUP) of Indian Council of Agricultural Research to prepare an agro-ecological region map of the country based on the parameters (rainfall, temperature, vegetation, potential evapo-transpiration) which form the-Bioclimate-and parameters (rainfall, potential evapotranspiration, soil storage) which constitutes the-Length of the Growing Period-(LGP) and parameters (soils and physiography) which form the-Soil or Land Scape. This article describes this classification.

Basis of Classification

The parameters taken for the classification agro-ecological regions are the characteristics of: physiographical features, soil characteristics, bio-climatic types and length of the growing period. These are explained one by one briefly.

Physiographic Features

The major physiographic regions in our country are the Mountainous region, Indo-Gangetic Alluvial Plains, Peninsular or Deccan Plateau and Coastal Plains. These are further divided into 19 broad and basic agro-ecological regions. Each region is represented by a capital letter of the English alpha-bets as given here.

Westem Himalayas

This includes Ladakh plateau, Kashmir Himalayas, Punjab Himalayas and Kumaun Himalayas of Uttar Pradesh and is symbolized by-A.

Central Himalayas

This includes mainly Nepal Himalayas, and foot-hills of V.P. Himalayas and is symbolized by-B.

Climate and its Influence on Crop Production

Climate change and agriculture are interrelated processes, both of which take place on a global scale. Global warming is projected to have significant impacts on conditions affecting agriculture, including temperature, carbon dioxide, glacial run-off, precipitation and the interaction of these elements. These conditions determine the carrying capacity of the biosphere to produce enough food for the human population and domesticated animals. The overall effect of climate change on agriculture will depend on the balance of these effects. Assessment of the effects of global climate changes on agriculture might help to properly anticipate and adapt farming to maximize agricultural production.

At the same time, agriculture has been shown to produce significant effects on climate change, primarily through the production and release of greenhouse gases such as carbon dioxide, methane, and nitrous oxide, but also by altering the Earth's land cover, which can change its ability to absorb or reflect heat and light, thus contributing to radiative forcing. Land use change such as deforestation and desertification, together with use of fossil fuels, are the major anthropogenic sources of carbon dioxide; agriculture itself is the major contributor to increasing methane and nitrous oxide concentrations in earth's atmosphere.

Impact of Climate Change on Agriculture

Despite technological advances, such as improved varieties, genetically modified organisms, and irrigation systems, weather is still a key factor in agricultural productivity, as well as soil properties and natural communities. The effect of climate on agriculture is related to variabilities in local climates rather than in global climate patterns. The Earth's average surface temperature has increased by 1 degree F in just over the last century. Consequently, agronomists consider any assessment has to be individually consider each local area. On the other hand, agricultural trade has grown in recent years, and now provides significant amounts of food, on a national level to major importing countries, as well as comfortable income to exporting ones. The international aspect of trade and security in terms of food implies the need to also consider the effects of climate change on a global scale.

A study published in *Science* suggest that, due to climate change, "southern Africa could lose more than 30% of its main crop, maize, by 2030. In South Asia losses of many regional staples, such as rice, millet and maize could top 10%". The 2001 IPCC Third Assessment Report concluded that the poorest countries would be hardest hit, with reductions in crop yields in most tropical and sub-tropical regions due to decreased water availability, and new or changed insect pest incidence. In Africa and Latin America many rainfed crops are near their maximum temperature tolerance, so that yields are likely to fall sharply for even small climate changes; falls in agricultural productivity of up to 30% over the 21st century are projected. Marine life and the fishing industry will also be severely affected in some places.

Climate change induced by increasing greenhouse gases is likely to affect crops differently from region to region. For example, average crop yield is expected to drop down to 50% in Pakistan according to the UKMO scenario whereas corn production in Europe is expected to grow up to 25% in optimum hydrologic conditions.

More favourable effects on yield tend to depend to a large extent on realization of the potentially beneficial effects of carbon dioxide on crop growth and increase of efficiency in water use. Decrease in potential yields is likely to be caused by shortening of the growing period, decrease in water availability and poor vernalization.

In the long run, the climatic change could affect agriculture in several ways :

- *productivity*, in terms of quantity and quality of crops
- *agricultural practices*, through changes of water use (irrigation) and agricultural inputs such as herbicides, insecticides and fertilizers
- *environmental effects*, in particular in relation of frequency and intensity of soil drainage (leading to nitrogen leaching), soil erosion, reduction of crop diversity
- *rural space*, through the loss and gain of cultivated lands, land speculation, land renunciation, and hydraulic amenities.
- *adaptation*, organisms may become more or less competitive, as well as humans may develop urgency to develop more competitive organisms, such as flood resistant or salt resistant varieties of rice.

They are large uncertainties to uncover, particularly because there is lack of information on many specific local regions, and include the uncertainties on magnitude of climate change, the effects of technological changes on productivity, global food demands, and the numerous possibilities of adaptation. Most agronomists believe that agricultural production will be mostly affected by the severity and pace of climate change, not so much by gradual trends in climate. If change is gradual, there may be enough time for biota adjustment.

Rapid climate change, however, could harm agriculture in many countries, especially those that are already suffering from rather poor soil and climate conditions, because there is less time for optimum natural selection and adaption.

Projections

The literature on key vulnerabilities to climate change. With low to medium confidence, they concluded that for about a 1 to 3°C global mean temperature increase (by 2100, relative to the 1990-2000 average level) there would be productivity decreases for some cereals in low latitudes, and productivity increases in high latitudes. With medium confidence, global production potential was predicted to:

- increase up to around 3°C,
- very likely decrease above about 3 to 4°C.

Most of the studies on global agriculture assessed by Schneider *et al.*. (2007:790) had not incorporated a number of critical factors, including changes in extreme events, or the spread of pests and diseases. Studies had also not considered the development of specific practices or technologies to aid adaptation.

Regional

- Africa:
    * Africa's geography makes it particularly vulnerable to climate change, and seventy per cent of the population rely on rain-fed agriculture for their livelihoods. Tanzania's official report on climate change suggests that the areas that usually get two rainfalls in the year will probably get more, and those that get only one rainy season will get far less. The net result is expected to be that 33% less maize—the country's staple crop—will be grown. Alongside other factors, regional climate change-in particular, reduced precipitation-is thought to have contributed to the conflict in Darfur. The combination of decades of drought, desertification and overpopulation are among the causes of the conflict, because the Baggara Arab nomads searching for water have to take their livestock further south, to land mainly occupied by farming peoples.
    * With high confidence, IPCC (2007:13) concluded that climate variability and change would severely compromise agricultural production and access to food.

- Asia: With medium confidence, IPCC (2007:13) projected that by the mid-21st century, in East and Southeast Asia, crop yields could increase up to 20%, while in Central and South Asia, yields could decrease by up to 30%. Taken together, the risk of hunger was projected to remain very high in several developing countries.
- Australia and New Zealand: Hennessy *et al.*. (2007:509) assessed the literature for this region. They concluded that without further adaptation to climate change, projected impacts would likely be substantial: By 2030, production from agriculture and forestry was projected to decline over much of southern and eastern Australia, and over parts of eastern New Zealand; In New Zealand, initial benefits were projected close to major rivers and in western and southern areas. Hennessy *et al.*. (2007:509) placed high confidence in these projections.
- Europe: With high confidence, IPCC (2007:14) projected that in Southern Europe, climate change would reduce crop productivity. In Central and Eastern Europe, forest productivity was expected to decline. In Northern Europe, the initial effect of climate change was projected to increase crop yields.
- Latin America: With high confidence, IPCC (2007:14) projected that in drier areas of Latin America, productivity of some important crops would decrease and livestock productivity decline, with adverse consequences for food security. In temperate zones, soybean yields were projected to increase.
- North America:
    * According to a paper by Deschenes and Greenstone (2006), predicted increases in temperature and precipitation will have virtually no effect on the most important crops in the US.
    * With high confidence, IPCC (2007:14-15) projected that over the first few decades of this century, moderate climate change would increase aggregate yields of rain-fed agriculture by 5–20%, but with important variability among regions. Major challenges were projected for crops that are near the warm end of their suitable range or which depend on highly utilized water resources.
- Polar regions (Arctic and Antarctic):
    * For the *Guardian* newspaper, Brown (2005) reported on how climate change had affected agriculture in Iceland.

Rising temperatures had made the widespread sowing of barley possible, which had been untenable twenty years ago. Some of the warming was due to a local (possibly temporary) effect via ocean currents from the Caribbean, which had also affected fish stocks.

* Anisimov *et al.*. (2007:655) assessed the literature for this region. With medium confidence, they concluded that the benefits of a less severe climate were dependent on local conditions. One of these benefits was judged to be increased agricultural and forestry opportunities.

- Small islands: In a literature assessment, Mimura *et al.*. (2007:689) concluded, with high confidence, that subsistence and commercial agriculture would very likely be adversely affected by climate change.

Shortage in Grain Production

Between 1996 and 2003, grain production has stabilized slightly over 1800 millions of tons. In 2000, 2001, 2002 and 2003, grain stocks have been dropping, resulting in a global grain harvest that was short of consumption by 93 millions of tons in 2003.

The Earth's average temperature has been rising since the late 1970s, with nine of the 10 warmest years on record occurring since 1995. In 2002, India and the United States suffered sharp harvest reductions because of record temperatures and drought. In 2003 Europe suffered very low rainfall throughout spring and summer, and a record level of heat damaged most crops from the United Kingdom and France in the Western Europe through Ukraine in the East. Bread prices have been rising in several countries in the region.

Poverty Impacts

Researchers at the Overseas Development Institute (ODI) have investigated the potential impacts climate change could have on agriculture, and how this would affect attempts at alleviating poverty in the developing world.

They argued that the effects from moderate climate change are likely to be mixed for developing countries. However, the vulnerability of the poor in developing countries to short term impacts from climate change, notably the increased frequency and severity of adverse weather events is likely to have a negative impact. This, they say, should be taken into account when defining agricultural policy.

Crop Development Models

Models for climate behaviour are frequently inconclusive. In order to further study effects of global warming on agriculture, other types of models, such as *crop development models*, *yield prediction*, quantities of *water or fertilizer consumed*, can be used. Such models condense the knowledge accumulated of the climate, soil, and effects observed of the results of various agricultural practices. They thus could make it possible to test strategies of adaptation to modifications of the environment. Because these models are necessarily simplifying natural conditions (often based on the assumption that weeds, disease and insect pests are controlled), it is not clear whether the results they give will have an *in-field* reality. However, some results are partly validated with an increasing number of experimental results.

Other models, such as *insect and disease development* models based on climate projections are also used (for example simulation of aphid reproduction or septoria (cereal fungal disease) development).

Scenarios are used in order to estimate climate changes effects on crop development and yield. Each scenario is defined as a set of meteorological variables, based on generally accepted projections. For example, many models are running simulations based on doubled carbon dioxide projections, temperatures raise ranging from 1°C up to 5°C, and with rainfall levels an increase or decrease of 20%. Other parameters may include humidity, wind, and solar activity. Scenarios of crop models are testing farm-level adaptation, such as sowing date shift, climate adapted species (vernalisation need, heat and cold resistance), irrigation and fertilizer adaptation, resistance to disease. Most developed models are about wheat, maize, rice and soybean.

Temperature Potential Effect on Growing Period

Duration of crop growth cycles are above all, related to temperature. An increase in temperature will speed up development. In the case of an annual crop, the duration between sowing and harvesting will shorten (for example, the duration in order to harvest corn could shorten between one and four weeks). The shortening of such a cycle could have an adverse effect on productivity because senescence would occur sooner.

Effect of Elevated Carbon Dioxide on Crops

Carbon dioxide is essential to plant growth. Rising CO_2 concentration in the atmosphere can have both positive and negative consequences. Increased CO_2 is expected to have positive physiological effects by increasing the rate of photosynthesis. Currently, the amount of carbon dioxide in the atmosphere is 380 parts per million. In comparison, the amount of oxygen is 210,000 ppm. This means that often plants may be starved of carbon dioxide, due to the enzyme that fixes CO_2, rubisco also fixes oxygen in the process of photorespiration. The effects of an increase in carbon dioxide would be higher on C3 crops (such as wheat) than on C4 crops (such as maize), because the former is more susceptible to carbon dioxide shortage. Studies have shown that increased CO_2 leads to fewer stomata developing on plants which leads to reduced water usage. Under optimum conditions of temperature and humidity, the yield increase could reach 36%, if the levels of carbon dioxide are doubled.

Further, few studies have looked at the impact of elevated carbon dioxide concentrations on whole farming systems. Most models study the relationship between CO_2 and productivity in isolation from other factors associated with climate change, such as an increased frequency of extreme weather events, seasonal shifts, and so on. In 2005, the Royal Society in London concluded that the purported benefits of elevated carbon dioxide concentrations are "likely to be far lower than previously estimated" when factors such as increasing ground-level ozone are taken into account."

Effect on Quality

According to the IPCC's TAR, "The importance of climate change impacts on grain and forage quality emerges from new research. For rice, the amylose content of the grain—a major determinant of cooking quality—is increased under elevated CO_2". Cooked rice grain from plants grown in high-CO_2 environments would be firmer than that from today's plants. However, concentrations of iron and zinc, which are important for human nutrition, would be lower. Moreover, the protein content of the grain decreases under combined increases of temperature and CO_2." Studies using FACE have shown that increases in CO_2 lead to decreased concentrations of micronutrients in crop plants. This may have knock-on effects on other parts of ecosystems as herbivores will need to eat more food to gain the same amount of protein.

Studies have shown that higher CO_2 levels lead to reduced plant uptake of nitrogen (and a smaller number showing the same for trace elements such as zinc) resulting in crops with lower nutritional value. This would primarily impact on populations in poorer countries less able to compensate by eating more food, more varied diets, or possibly taking supplements. Reduced nitrogen content in grazing plants has also been shown to reduce animal productivity in sheep, which depend on microbes in their gut to digest plants, which in turn depend on nitrogen intake.

Agricultural Surfaces and Climate Changes

Climate change may increase the amount of arable land in high-latitude region by reduction of the amount of frozen lands. A 2005 study reports that temperature in siberia has increased three degree Celsius in average since 1960 (much more than the rest of the world). However, reports about the impact of global warming on Russian agriculture indicate conflicting probable effects : while they expect a northward extension of farmable lands, they also warn of possible productivity losses and increased risk of drought. Sea levels are expected to get up to one meter higher by 2100, though this projection is disputed. A rise in the sea level would result in an agricultural land loss, in particular in areas such as South East Asia. Erosion, submergence of shorelines, salinity of the water table due to the increased sea levels, could mainly affect agriculture through inundation of low-lying lands.

Low lying areas such as Bangladesh, India and Vietnam will experience major loss of rice crop if sea levels are expected to rise by the end of the century. Vietnam for example relies heavily on its southern tip, where the Mekong Delta lies, for rice planting. Any rise in sea level of no more than a meter will drown several km^2. of rice paddies, rendering Vietnam incapable of producing its main staple and export of rice.

Erosion and Fertility

The warmer atmospheric temperatures observed over the past decades are expected to lead to a more vigorous hydrological cycle, including more extreme rainfall events. Erosion and soil degradation is more likely to occur. Soil fertility would also be affected by global warming. However, because the ratio of carbon to nitrogen is a constant, a doubling of carbon is likely to imply a higher storage of nitrogen in soils as nitrates, thus providing higher fertilizing elements for plants, providing better yields. The average needs for nitrogen could decrease, and give the opportunity of changing often costly fertilisation strategies.

Due to the extremes of climate that would result, the increase in precipitations would probably result in greater risks of erosion, whilst at the same time providing soil with better hydration, according to the intensity of the rain. The possible evolution of the organic matter in the soil is a highly contested issue: while the increase in the temperature would induce a greater rate in the production of minerals, lessening the soil organic matter content, the atmospheric CO_2 concentration would tend to increase it.

The warmer atmospheric temperatures observed over the past decades are expected to lead to a more vigorous hydrological cycle, including more extreme rainfall events. In 1998 Karl and Knight reported that from 1910 to 1996 total precipitation over the contiguous U.S. increased, and that 53% of the increase came from the upper 10% of precipitation events (the most intense precipitation). The percent of precipitation coming from days of precipitation in excess of 50 mm has also increased significantly.

Studies on soil erosion suggest that increased rainfall amounts and intensities will lead to greater rates of erosion. Thus, if rainfall amounts and intensities increase in many parts of the world as expected, erosion will also increase, unless amelioration measures are taken. Soil erosion rates are expected to change in response to changes in climate for a variety of reasons.

The most direct is the change in the erosive power of rainfall. Other reasons include: a) changes in plant canopy caused by shifts in plant biomass production associated with moisture regime; b) changes in litter cover on the ground caused by changes in both plant residue decomposition rates driven by temperature and moisture dependent soil microbial activity as well as plant biomass production rates; c) changes in soil moisture due to shifting precipitation regimes and evapo-transpiration rates, which changes infiltration and runoff ratios; d) soil erodibility changes due to decrease in soil organic matter concentrations in soils that lead to a soil structure that is more susceptible to erosion and increased runoff due to increased soil surface sealing and crusting; e) a shift of winter precipitation from non-erosive snow to erosive rainfall due to increasing winter temperatures; f) melting of permafrost, which induces an erodible soil state from a previously non-erodible one; and g) shifts in land use made necessary to accommodate new climatic regimes.

Effects of Global Climate Change on Pests, Diseases and Weeds

A very important point to consider is that weeds would undergo the same acceleration of cycle as cultivated crops, and would also benefit from carbonaceous fertilization. Since most weeds are C3 plants, they are likely to compete even more than now against C4 crops such as tomatoes. However, on the other hand, some results make it possible to think that weedkillers could gain in effectiveness with the temperature increase.

Global warming would cause an increase in rainfall in some areas, which would lead to an increase of atmospheric humidity and the duration of the wet seasons. Combined with higher temperatures, these could favour the development of fungal diseases. Similarly, because of higher temperatures and humidity, there could be an increased pressure from insects and disease vectors.

Glacier Retreat and Disappearance

The continued retreat of glaciers will have a number of different quantitative impacts. In areas that are heavily dependent on water runoff from glaciers that melt during the warmer summer months, a continuation of the current retreat will eventually deplete the glacial ice and substantially reduce or eliminate runoff. A reduction in runoff will affect the ability to irrigate crops and will reduce summer stream flows necessary to keep dams and reservoirs replenished.

Approximately 2.4 billion people live in the drainage basin of the Himalayan rivers. India, China, Pakistan, Afghanistan, Bangladesh, Nepal and Myanmar could experience floods followed by severe droughts in coming decades. In India alone, the Ganges provides water for drinking and farming for more than 500 million people. The west coast of North America, which gets much of its water from glaciers in mountain ranges such as the Rocky Mountains and Sierra Nevada, also would be affected.

Ozone and UV-B

Some scientists think agriculture could be affected by any decrease in stratospheric ozone, which could increase biologically dangerous ultraviolet radiation B. Excess ultraviolet radiation B can directly effect plant physiology and cause massive amounts of mutations, and indirectly through changed pollinator behaviour, though such changes are simple to quantify. However, it has not yet been ascertained whether an increase in greenhouse gases would decrease stratospheric ozone levels. In addition, a possible effect of rising temperatures is significantly higher levels of ground-level ozone, which would substantially lower yields.

ENSO Effects on Agriculture

ENSO (El Niño Southern Oscillation) will affect monsoon patterns more intensely in the future as climate change warms up the ocean's water. Crops that lie on the equatorial belt or under the tropical Walker circulation, such as rice, will be affected by varying monsoon patterns and more unpredictable weather. Scheduled planting and harvesting based on weather patterns will become less effective.

Areas such as Indonesia where the main crop consists of rice will be more vulnerable to the increased intensity of ENSO effects in the future of climate change. University of Washington professor, David Battisti, researched the effects of future ENSO patterns on the Indonesian rice agriculture using [IPCC]'s 2007 annual report and 20 different logistical models mapping out climate factors such as wind pressure, sea-level, and humidity, and found that rice harvest will experience a decrease in yield. Bali and Java, which holds 55% of the rice yields in Indonesia, will be likely to experience 9-10% probably of delayed monsoon patterns, which prolongs the hungry season. Normal planting of rice crops begin in October and harevest by January. However, as climate change affects ENSO and consequently delays planting, harvesting will be late and in drier conditions, resulting in less potential yields.

Environmental Issues with Agriculture

There are numerous environmental issues with the various practices of agriculture.

Climate Change

Climate change and agriculture are interrelated processes, both of which take place on a global scale. Global warming is projected to have significant impacts on conditions affecting agriculture, including temperature, precipitation and glacial run-off. These conditions determine the carrying capacity of the biosphere to produce enough food for the human population and domesticated animals. Rising carbon dioxide levels would also have effects, both detrimental and beneficial, on crop yields. The overall effect of climate change on agriculture will depend on the balance of these effects. Assessment of the effects of global climate changes on agriculture might help to properly anticipate and adapt farming to maximize agricultural production.

At the same time, agriculture has been shown to produce significant effects on climate change, primarily through the production and release of greenhouse gases such as carbon dioxide, methane, and nitrous oxide, but also by altering the Earth's land cover, which can change its ability to absorb or reflect heat and light, thus contributing to radiative forcing. Land use change such as deforestation and desertification, together with use of fossil fuels, are the major anthropogenic sources of carbon dioxide; agriculture itself is the major contributor to increasing methane and nitrous oxide concentrations in earth's atmosphere.

Deforestation

One of the causes of deforestation is to clear land for pasture or crops. According to British environmentalist Norman Myers, 5% of deforestation is due to cattle ranching, 19% due to over-heavy logging, 22% due to the growing sector of palm oil plantations, and 54% due to slash-and-burn farming.

In 2000 the United Nations Food and Agriculture Organization (FAO) found that "the role of population dynamics in a local setting may vary from decisive to negligible," and that deforestation can result from "a combination of population pressure and stagnating economic, social and technological conditions."

Genetic Engineering

Genetic engineering has caused controversies.

Seed contamination is problematic.

Intensive Farming

Intensive farming alters the environment in many ways. Some of the disadvantages of this method of farming include:

- Limits or destroys the natural habitat of most wildlife, and leads to soil erosion
- Use of fertilizers can alter the biology of rivers and lakes.
- Pesticides generally kill useful insects as well as those that destroy crops
- Generally not sustainable-often results in desertification or, in a worst case scenario, land that is so poisonous and eroded that nothing else will grow
- Requires large amounts of energy input to produce, transport, and apply chemical fertilizers/pesticides
- Use of chemicals on fields creates run-off, excess runs off into rivers and lakes causing pollution

- Use of pesticides have numerous negative health effects in workers who apply them, people that live nearby the area of application or downstream/downwind from it, and consumers who eat the pesticides which remain on their food

Irrigation

Irrigation can lead to a number of problems :

- Depletion of underground aquifers through overdrafting.
- Ground subsidence.
    - o Groundwater recharge-an ecological restoration, mitigation, and remediation technique.
- Underirrigation gives poor soil salinity control which leads to increased soil salinity with consequent build up of toxic salts on soil surface in areas with high evaporation. This requires either leaching to remove these salts and a method of drainage to carry the salts away.
- Overirrigation because of poor distribution uniformity or management wastes water, chemicals, and may lead to water pollution.
- Deep drainage (from over-irrigation) may result in rising water tables which in some instances will lead to problems of irrigation salinity requiring watertable control by some form of subsurface land drainage.
- Irrigation with saline or high-sodium water may damage soil structure owing to the formation of alkaline soil.
- Runoff causing surface water and groundwater-aquifer hydrologic cycle water pollution.
    - o Bioretention-an ecological restoration, mitigation, and remediation technique.

Pollutants

A wide range of agricultural chemicals are used and some become pollutants through use, misuse, or ignorance.

- Pesticide drift
    - o soil contamination
    - o groundwater and water pollution
    - o air pollution *spray drift*
- Pesticides, especially those based on organochloride
- Pesticide residue in foods

- Pesticide toxicity to bees
    - o List of crop plants pollinated by bees
    - o Pollination management
- Bioremediation.

Waste

Plasticulture, the use of plastic materials in agriculture, raises problems around how to carry out the recycling of agricultural plastics.

Issues by Region

- Hedgerow removal in the United Kingdom.
- Soil salinisation, especially in Australia.
- Phosphate mining in Nauru
- Methane emissions from livestock in New Zealand.
- Some environmentalists attribute the hypoxic zone in the Gulf of Mexico as being encouraged by nitrogen fertilization of the algae bloom.

Sustainable Agriculture

The exponential population increase in recent decades has increased the practice of agricultural land conversion to meet demand for food which in turn has increased the effects on the environment. The global population is still increasing and will eventually stabilise, as some critics doubt that food production, due to lower yields from global warming, can support the global population.

Organic farming is a multifaceted sustainable agriculturc set of practices that can have a lower impact on the environment

Other specific methods include: permaculture; and biodynamic agriculture which incorporating a spiritual element.

- Category: Sustainable agriculture
- Biological pest control.

8

Scope of Medicinal and Aromatic Plants in Arid Region

Diversification of agriculture refers to the shift from the regional dominance of one crop to regional production of a number of crops, to meet ever increasing demand for cereals, pulses, vegetables, fruits, oilseeds, fibres, fodder and grasses, fuel, etc. It aims to improve soil health and a dynamic equilibrium of the agro-ecosystem. Crop diversification takes into account the economic returns from different value-added crops. It is different from the concept of multiple cropping or succession planting in which multiple crops are planted in succession over the course of a growing season.

Moreover, it implies the use of environmental and human resources to grow a mix of crops with complementary marketing opportunities, and it implies a shifting of resources from low value crops to high value crops, usually intended for human consumption such as fresh market fruits and vegetables. With globalization of the market, crop diversification in agriculture means to increase the total crop productivity in terms of quality, quantity and monetary value under specific, diverse agro-climatic situations world-wide.

There are two approaches to crop diversification in agriculture. First is horizontal diversification, which is the primary approach to crop diversification in production agriculture. Here, diversification takes place through crop intensification by adding new high-value crops to existing cropping systems as a way to improve the overall productivity of a farm or region's farming economy. The second is the vertical diversification approach in which farmers and others add value to products through processing, regional branding, packaging, merchandising, or other efforts to enhance the product. Opportunities for crop diversification vary depending on risks, opportunities and the feasibility of proposed changes within a socio-economic and agro-economic context.

Crop diversification may occur as a result of government policies. The "Technology Mission on Oilseeds", "Spices Development Board", "Coconut Development Board" etc. are examples where the Indian government created policies to thrust change upon farmers and the food supply chain at large as a way to promote crop diversity. Crop diversification is the outcome of several interactive effects of many factors:

1. Environmental factors including irrigation, rainfall, temperature, and soil fertility.
2. Technology-related factors including seeds, fertilizers and water technologies, but also those related to marketing, harvest, storage, agro-processing, distribution, logistics, etc.
3. Household-related factors including regional food traditions, fodder and fuel as well as the labor and investment capacity of farm people and their communities.
4. Price-related factors including output and input prices as well as national and international trade policies and other economic policies that affect the prices either directly or indirectly.
5. Institutional and Infrastructure-related factors including farm size, location and tenancy arrangements, research, in-field technical support, marketing systems and government regulating policies, etc.

All these five factors are interrelated. The adoption of crop technologies is commonly assumed to be influenced primarily by resource-related factors when institutional and infrastructure factors can play as much or more of a role in their adoption.

Area Expansion Problems under Rice and Wheat Crops

Scaling up production area poses several new problems of significance such as:

1. Excessive use of groundwater leading to poor water use efficiency and depletion of groundwater.
2. Deterioration of soil health or soil fertility.
3. Multiple infestations of weed flora, insect pests and diseases.

4. Indiscriminate use of energy such as chemical, electricity or disease, etc.
5. Reduction in the availability of other protective food and high value crops.
6. Pollution of agro-ecosystems.

On the other hand, crop diversification has potential as an economic driver in agricultural regions. It may prove to be of paramount importance in meeting challenges that arise from a post-green revolution scenario. In view of shrinkage of agricultural land and operational holdings due to expansion of urban centers, changes in consumer food habits, exponential population growth rate, farmers are pressured to include or substitute additional crops in to the cropping system.

Major Driving Forces for Crop Diversification

The major driving forces for crop diversification are:

1. Increasing income on small farm holdings.
2. Withstanding price fluctuation.
3. Mitigating ill-effects of aberrant weather.
4. Balancing food demand.
5. Improving fodder for livestock animals.
6. Conservation of natural resources (soil, water, etc.).
7. Minimizing environmental pollution.
8. Reducing dependence on off-farm inputs.
9. Decreasing insect pests, diseases and weed problems.
10. Increasing community Food security

Indian agriculture is characterized by a dominance of small and marginal farmers (almost 68 per cent) who suffer as a result of difficult socio-economic conditions. 75 per cent of the farm holdings are below 2 hectares, and a large portion of rural people subsist as small holders. Income from these farms can not be raised up to the desired level to sufficiently alleviate poverty in the countryside unless existing crop production systems are diversified through inclusion of high value horticultural and arable crops. Furthermore, increased dependence on one or two major cereal crops (wheat, rice, etc.) witnessed after the green revolution makes the farming economy vulnerable to price fluctuation arising due to demand-supply or export-import equations especially after the WTO began influencing markets.

Crop diversification on the other hand, can better tolerate the ups and downs in the market value of farm products and may ensure economic stability for farming families of the country. The adverse effects of aberrant weather, such as erratic and scanty rainfall and drought are very common in a vast area in agricultural production of the country. Incidence of flood in one part of the country and drought in the other part is a very frequent phenomenon in India. Under these aberrant weather situations, dependence on one or two major cereals (rice, wheat, etc.) is always risky.

Hence, crop diversification through substitution of one crop or mixed cropping/inter-cropping may be a useful tool to mitigate problems associated with aberrant weather to some extent, especially in the arid and semi-arid drought-prone/dryland areas.

Immediate Need

Crop diversification in agriculture in India is taking place vertically or horizontally, mostly due to market forces and occasionally due to domestic needs. Where there are concerns regarding land and water use and quality, there is immediate need to consider:

1. Processing of farm produce into value added products offers scope for employment in non-farm works such as distillation of active ingredients from medicinal and aromatic plants (herbal products), scope of industrialization in agriculture for sugar, paper board manufacturing, etc. to increase employment in rural areas.
2. There is a need to generate place-based approaches for diverse farming situations under various socio-economic conditions, domestic needs, market infrastructure, input supply, etc.
3. The research on crop diversification is best done in a farmer-participatory mode in which a multi-disciplinary team of scientists involves farmers from project planning through arriving at conclusions.
4. A concept of sustainable productivity for each unit of land and water through crop diversification needs to be fostered.
5. There is need for promoting co-operatives in rural areas to solve micro-level and location-specific problems.
6. Major thrust should be given on horticulture (vegetables, fruits, flowers, spices, etc.) and animal husbandry (dairying, poultry, goatery, piggery, duckery, etc.) to support a vigorous and expanding export market, balanced with supplying local markets with affordable, healthy food.

7. Strengthening food processing and other value-added industries in rural areas is a means to provide employment to rural youth.
8. There is need to develop rural infrastructure such as roads, markets, medical and educational facilities in the villages with efficient utilization of local resources for farming community in a more pragmatic way.
9. Crop diversification provides efficient use of farm inputs and contributes to a strong rural economy.
10. Alternate cropping systems and farm enterprise diversification are most important for generating higher income, employment and protecting the environment.
11. There are numerous opportuntiies to adopt subsidiary occupations to the rice-wheat cropping systems common in India. These include vegetable farming, fruit cultivation, floriculture, medicinal and aromatic plants cultivation, mushroom farming, dairying, piggery, goatery, poultry and duckery, fishery or aquaculture, bee-keeping, agroforestry, biodiesel farming with Jatropha Curcas (veranda), palm, neem, Karanja, etc. to provide ample scope for diversification of rice-wheat cropping system in north-western and south India and north-eastern states.
12. Enterprise diversification generates more income and rural employment year round.

Medicinal plants are the local heritage with global importance, World is endowed with a rich wealth of medicinal plants. Herbs have always been the principal form of medicine in India and presently they are becoming popular throughout the developed world, as people strive to stay healthy in the face of chronic stress and pollution, and to treat illness with medicines that work in concert with the body's own defense. People in Europe, North America and Australia are consulting trained herbal professionals and are using the plant medicines. Medicinal plants also play an important role in the lives of rural people, particularly in remote parts of developing countries with few health facilities.

The variety and sheer number of plants with therapeutic properties is quite astonishing. It is estimated that around 70,000 plant species, from lichens to towering trees, have been used at one time or another for medicinal purposes. The herbs provide the starting material for the isolation or synthesis of conventional drugs.

In Ayurveda about 2,000 plant species are considered to have medicinal value, while the Chinese *Pharmacopoeia* lists over 5,700 traditional medicines, most of which are of plant origin. About 500 herbs are still employed within conventional medicine, although whole plants are rarely used.

In India, medicinal plants have made a good contribution to the development of ancient Indian *Material Medica.* One of the earliest treatises on Indian medicine, the *Charak Samhita* (1000 B.C.), records the use of over 340 drugs of vegetable origin. Most of these continue to be gathered from wild plants to meet the demand of the medical profession. Thus, despite the rich heritage of knowledge on the use of plant drugs, little attention had been paid to grow them as field crops in the country till the latter part of the nineteenth century.

Medicinal plants help in alleviating human suffering. These plants "are being integrated to the field of foods as additives, beverages and cosmetics. They are widely used as sweeteners, as biters, as spices, as natural colouring agent and as insecticides. Mass selection recurrent selection, hybridization, clonal selection mutation and biotechnology are some of major techniques at their use for many proven medicinal plants. There are still several constraints.

During the past one century there has been a rapid extension of the allopathic system of medical treatment in India. It generated commercial demand for pharmacopoeial drugs and their products in India. Efforts have been made to introduce many of these drug plants to farmers. Several research institutes have undertaken studies on the cultivation practices of medicinal plants, which were found suitable and remunerative for commercial cultivation. The agronomic practices for growing poppy, *isabgol, senna,* cinchona, ipecac, belladonna, ergot and few others have been developed and there is now localized cultivation of these medicinal plants commercially.

Medicinal plants have curative properties due to the presence of various complex chemical substances of different composition, which are found as secondary plant metabolites in one or more parts of these plants. These plant metabolites, according to their composition, are grouped as alkaloids, glycosides, corticosteroids, essential oils, etc. The alkaloids form the largest group, which includes morphine and *codeine* (Poppy), strychnine and *brucine (Nux vomica), quinine (Cinchona),* ergotamine *(Ergot), hyocyamine (Belladonna)* scolapomine *(Datura),* emetine *(Ipecac),* cocaine *(Coco),* ephedrine *(Ephedra),* reserpine *(Rauwolfia),* caffeine (Tea dust), aconitine *(Aconite),* vascine *(Vasaca)* santonin *(Artemisia),* lobelin *(Lobelia)* and a large number of others. Glycosides form another important group represented by digoxin *(Foxglove),* stropanthin *(Strophanthus),* glycyrrhizin *(Liquorice),* barbolin *(Aloe),* sannocides *(Senna),* etc. Corticosteroids have come into sannocides *(Senna),* etc. Corticosteroids have come into prominence recently and diosgenin *(Dioscorea),* solasodin *(Solanum* sp.), etc. now command a large world demand. Some essential oils such as those of valerian *kutch* and peppermint also possess medicating properties and are used in the pharmaceutical industry. However, it should be stated in all fairness that our knowledge of the genetic and physiological make-up of most of the medicinal plants is poor and we know still less about the biosynthetic pathways leading to the formation of active constituents for which these crops are valued.

Medicinal and aromatic plants are found in forest areas throughout South Asia, from the plains to the high Himalayas, with the greatest concentration in the tropical and subtropical belts and arid region of Thar desert. India recognizes more than 2,500 plant species as having medicinal value, Sri Lanka about 1,400, and Nepal around 700. Some of these, found at high altitudes in particularly stressful environments, grow very slowly and cannot live elsewhere. Others are more broadly distributed and adapt more easily to different ecological conditions.

During the past decade, a dramatic increase in exports of medicinal plants attests to worldwide interest in these products as well as in traditional health systems. In the last 10 years, for example, India's exports of medicinal plants have trebled. But with most of these plants being taken from the wild, hundreds of species are now threatened with extinction because of over-harvesting, destructive collection techniques, and conversion of habitats to crop-based agriculture. For instance, the small coniferous Himalayan yew *(Taxus baccata)* has recently become a heavily traded species. Similarly, *senna* is being grown extensively in arid region of India.

The pharmaceutical industries have made massive investment on pharmacological, clinical and chemical researches all over the world in past five decades. Efforts have been made to discover still more potent plant drugs. In fact, a few new drug plants have successfully been passed the tests of commercial screening. The benefits of these efforts would reach to the masses in future if farmers initiate commercial cultivation of medicinal plants. In fact, agricultural studies on medicinal plants, by its very nature, demand an equally large investment and higher priority. India, in particular, has a big scope for the development of pharmaceutical and phytochemical industry.

The subcontinent, India is blessed with varieties of aromatic and medicinal plants. The agroclimatic conditions and rainfall favouring this bio-availability. More than 7,500 species of medicinal plants are grown in India. Owing to this India is considered as the botanical garden of the world and treasure house of the biodiversity. Ayurveda, our indigenous system of health care is accepted everywhere especially abroad. Vedas and other ancient scriptures give cleanout evidences of using herbs and medicinal plants. Ayurveda alone describes about 2000 species of plants, which constitutes more than 10,000 formulations.

Over the past 10 years there has been a considerable interest in the use of herbal medicines in the world. Regarding the export of medicinal plants India's contribution to the international market is comparatively very low. Utilizing our biodiversity and proper planning, Indian products can very well enter the overseas markets. This can be achieved only through proper development of medicinal plants, standardization of the extracts and keeping the quality. WHO has recognized the effectiveness of traditional system of medicine and its safety.

The Indian Pharmacopoeia (1966) recognized eighty five drug plants whose ingredients are used in various pharmaceutical preparations. The text is however; confine to a few important commercially grown medicinal plants whose cultivation deserves priority in out national economy.

According to R. B. S. Rawat and R. C. Uniyal, National Medicinal Plants Board Department of ISM&H *(Agrobios News Letter, Vol. 1, No. 8, January 2003)* the use of medicinal plants is as old as human civilization. India has a glorious tradition of health care system based on plants, which dates back to *Vedic* era. In *Rig Veda* which is the oldest known repository of human knowledge and wisdom (4500-2500 B.C.) mentions about hundred medicinal plants used by the Aryans while in *Atharva Veda* (2500-2000 B.C.) elaborate description of medicinal plants are given. Later in *Samhita* period the science of medicine systematically organized with clear concept and theories based on the treatises the *Charak Samhita* - 2000 B. C, *Sushruta Samhita* - 1000-800 B.C. Besides this there are other works on Ayurveda and medicinal plants by Nagarjun, Chakradatta, Sharangadhar and Bangasen - 1000-500 B.C. Vaghabhatta Junior - 800 A. D. complied most of the books on Ayurveda and wrote Ashtanga Hariday Samhita.

They further indicated that India is bestowed with unique diversity in culture and natural vegetation exhibiting rich plant diversity. It has all known types of agro-climatic, ecologic and edaphic conditions. It also have unique biogeographical positions having all known types of eco-systems ranging from coldest place, the dry cold desert of Ladakh (Nubra Valley with - 57°C), to temperate, alpine and sub-tropical regions of north-west and trans-Himalayas; rain forests with high rainfall; wet evergreen humid tropics of western ghats and arid and semi-arid regions of peninsular India; dry desert conditions of Rajasthan and Gujarat to the tidal mangroves of Sunderban.

It harbors 17500 flowering plants out of which 2000 plants are used in various classical system of medicine like Ayurveda, Siddha and Unani. The tribal and other communities use about 8000 species of wild plants as traditional medicine. The drugs used in ISM are 90% based on plant material and are considered to be safe, cost effective and with minimal or no side effects when genuine ingredients are used.

Medicinal plants are living and irreparable resource, which is exhaustible if over used and sustainable if used with care and wisdom. The importance of medicinal plants has been overlooked in the past. However, at present medicinal plants are looked upon not only as a source of affordable health care but also as a source of income. According to WHO report, over 80% of the world population relies on traditional medicine largely plant based for their primary healthcare needs.

The forest areas have been the traditional source of medicinal plants and herbs. The position cannot be sustained much further because on the one hand the areas under forests have been steadily shrinking and on the other the requirement of medicinal plants and herbs has increased steeply. This has resulted in unscientific and over exploitation of medicinal plants in the forests. One indication of the scarcity of some medicinal plants is their steep prices. The Ministry of Environment and Forests have already banned 29 endangered species of medicinal plans from their natural habitat. While medicinal plants are being utilized in the preparation of a number of modern drugs, there is a new trend worldwide of using natural products. Besides medicinal values, Pharmaceuticals, herbal food supplements, toiletries and cosmetics are growing in consumption in the international market.

Diversity of Horticultural Crops in North Eastern Region

The North East region has its own unique combination of living species, habitats and ecosystems, which together make up its diversity rich resource. While speaking strictly about plant diversity, two regions of the country are termed as hot spots. These are – Western Ghats and the North Eastern hill regions.

In all living organisms the species is the single most useful unit to use in diversity assessment. Species richness and the relative abundance of different species is another criterion to measure the degree of diversity. The number of endemic species also reflects into account while assessing the richness of diversity.

The North Eastern region of India comprising eight states namely Arunachal Pradesh, Assam, Manipur, Meghalaya, Mizoram, Nagaland, Tripura and Sikkim has vast physiographical variations, which have been represented in 6 agro climatic zones.

North East region is one of the richest reservoir of genetic variability and diversity of different crops i.e various kinds of fruits, different vegetables, spices, ornamental plants and also medicinal & aromatic plants. The Diversity for horticultural crops of this region has mainly been managed by local farmers, often women.

Considerable diversity exists among the regional horticultural species including variation in plant type, morphological and physiological characteristics, reactions to diseases and pests, adaptability and distribution. Apart from the nutritional value, many regional horticultural crops are used for medicinal purposes, income generating and poverty alleviation programmes in the rural areas.

Exploration and Distribution

Fruits

Citrus: Being the home of several citrus species, rich genetic diversity occurs in the region. Bhattacharya and Dutta (1956) described 17 citrus species, their 52 cultivars and a few probable natural hybrids from this region. In lemon alone, as many as 32 strains are available. The species *C. limon, C. medica, C. jambhiri, C. ichangensis, C. latipes, C. macroptera, C. assamensis, C. indica* and *C. aurantium* are considered indigenous to this region. The Indian wild orange *C. indiaca* is found in the Naga hills and Meghalaya.

Banana and Mango: Maximum genetic variability of *Musa acuminata* and *M. balbisiana* occurs in NE India. *M. flaviflora* is localized to Manipur and Meghalaya. There are other species found in Sikkim and Khasi Hills, which need systematic collection and conservation. Some native *Mangifera spp.* are found in Tripura, Manipur, Mizoram and South Assam. Wild form of *M. indica* and its allied species *M. sylvetica* occur in Arunachal Pradesh, *M. khasiana* and *M. pentandra* in Assam. Temperate fruits: Rich diversity occurs in Pyrus, Rubus, Ribes and Prunus. The Shillong plateau of Khasi hills in Meghalaya has many Prunus species such as *P. napalensis*, *P. undulata* and *P. cerasoides*. *Pyrus pyrifolia* var. cubha makai (*P. serotina* Red) is grown semi commercially in Meghalaya, Manipur and other places. Wild kiwi (*Actinidia callosa* and *A. stragosa*) is found growing in the natural forests of Arunachal Pradesh and Sikkim. Tropical and sub tropical fruits: A large number of other tropical and subtropical fruits belonging to the genera Garcinia, Artocarpus, Phyllanthus, Annona, Averrhoa, Persia, Aegle, Passiflora, etc., arc found growing wild in the region. One of the indigenous fruits that requires attention is jackfruit, which grows abundantly in Tripura, Assam and Meghalaya with a large number of cultivars.

Under Utilized Fruits: Of the 300 edible plant species found in the NE region, some of them are really worth consuming by various ethnic groups of tribals. Two species of *Elaegnus*, viz *E. latifolia* and *E. pyriformis* are known to be grown in NE region (Pandey, 2002). It is quite common in Sibsagar (Dikho valley), Naga hills, Khasi and Jaintia hills. *Docynia indica* and *D. hookeriana* are commonly found in the region. *Pyrus pashia* is a medium sized deciduous fruit tree also found in NE region.

Tomato and Chili: Tomato is an introduced crop roughly in 18th century and most of the introductions are bred varieties, which have adapted to this region (Seshadri and Srivastava, 2002).

Germplasm of wild species of tomato *L. pimpinellifolium,* has been found in NE region. Chilies are usually grown in warm to hot and humid climate in Manipur, Mizoram, Meghalaya, Nagaland, Tripura, and Arunachal Pradesh in that order with respect to area under the crop.

Due to the long history of cultivation, outcrossing nature and popularity of the crop, large genetic diversity including local landraces have evolved. In hot chili great range of variability for several attributes (fruit shape, size, colour and bearing habit and semi-perennial, perennial and pungency) occur throughout the North Eastern region.

Cucurbitaceous Vegetables: Consisting of more than 15 kinds, these vegetables are grown and consumed within the region. In North East many species of cucurbits are found as vegetables and fruits; these include *Cucurbita*, *Momordica*, *Luffa* and several lesser known cucurbitaceous crops.

Pumpkin varieties abound in number with variation in fruit size, fruit skin, flesh colour thickness, sweetness, etc. The wild species *Cucumis hardwickii*, the likely progenitor of cultivated cucumber, is found growing in natural habitats in the foothill of Himalayas and NE region particularly in Meghalaya. *C. sativus* var. *sativus* is cultivated all North Eastern region in tropical and subtropical conditions. Among gourds, in North Eastern region maximum variability has been recorded for bottle gourd in fruit shape and size. The NE region has rich diversity in genetic resources of ridge gourd (*L. acutangula*) and sponge gourd (*L. cylindrica*). Small as well as large sized forms of bitter gourd are also available.

Crucifers: These are essentially cruciferous vegetables namely cauliflower, cabbage, knolkhol, etc., introduced from the days of East India Company in 14th-15th century when European traders visited this region (Seshadri and Srivastava, 2002). Variability in Indian cauliflower widely exists in Assam and Meghalaya.

Leguminous Vegetables: A wide variability of French bean, cowpea and Indian bean is found in the various parts of the region. In French bean, climbing or pole type is popular among the tribals since it is used for mix cropping with maize, the stem of which act as the support for the bean. One of the interesting species of Vigna namely *V. vexillata* is grown by the tribals of Tripura. It is a legume cum tuber crop with much variation in edible tubers (Arora and Pandey, 1996). Sword bean (*Canavalia ensiformis* (L) DC) of papilionaceae family is also cultivated on limited scale in the North Eastern region (CSIR, 1950). Winged bean is confined in humid sub tropical parts of NE region.

Leafy Vegetables: The important leafy vegetables include lai (*Brassica juncea*), lafa (*Malva verticillata*), palak (*Spinacea oleracea*). In addition to these a wide variety of indigenous leafy vegetables are also available. These are amaranth (*Amaranthus spp*), puroi sag (*Vasella rubra* and *B. alba*), sorrel (*Rumex rasicarius*), etc. Other indigenous leafy vegetables used occasionally are jilmilsag (*Chenopodium album*) and Kalmou sag (*Ipomea reptans*). *Amaranthus viridis, A. lividus, A. retroflexus* and *A. spinosus* are important leafy types grown in North East India (Sarma, 2001).

Tuber and Rhizomatus Crops: Based on the colour and the skin to broad types of sweet potato are grown in the region. These are the white skinned and the red skinned variety. A number of dioscorea species *alata, bulbifera, brevipetiolata, esculenta, hamiltonii, hispida, kamaonensis, nummularia, pentaphylla, puber* and *quinata* were recorded in the Region. *D. hamiltonii* occurs in humid forests of NE hills (Sarma, 2001). Varieties of tapioca or cassava like M-4 and hybrids H-97, H-165, H-226 have also been cultivated to some extent. In colocasia also there is a wide variability even with in one species such as *Colocasia esculenta* (Sarma, 2001).

Lesser-known Vegetables: In addition to the above there are a large number of indigenous vegetable crops that are used particularly by the tribal population. Maximum such varieties are available in Arunachal Pradesh. Tree bean (*Parkia roxburghii* G. Don.) is one of the most common of multipurpose tree species in Manipur and Mizoram (Kumar *et al.*, 2002). In the hilly areas tree tomato ENVIS Centre, GBPIHED 6 (*Cyphomandra betacca*), a perennial shrub producing red tomato like vegetables, is also grown and used as such. It is grown as backyard venture crop in Meghalaya (Thakur *et al.*, 1988). Another vegetable tree, growing in the lower altitude zones and popular among the people, is drum stick or horse reddish locally called Sajina (*Moringa oleifera*). Cho-Cho (*Sechium edule*), a native of tropical America, is a very popular vegetable in the region commonly called squash and grows abundantly without much care and attention.

Flemingia vestita known as Sohphlong, is consumed raw. It is a weak climbing/trailing type, under ground tubers, distributed the humid to sub tropical regions of NE India upto 1500m (Sarma, 2001). Kakrol (*Momordica cochinchinesis*) and kartoli (*M. dioica*) are widely spread in Assam, the Garo hills of Meghalaya (Ram *et al.,* 2002).

Spices: A wide range of variability has been found in ginger and turmeric. Lakadong the local variety of turmeric is grown in Jowai area of Meghalaya. Poona and Thinglaidon performed better in most of the states except Tripura and Sikkim indicating suitability for large scale cultivation for green ginger production. Nadia variety of ginger is well adopted in North East region. A type of ginger having rhizomes with bluish black tinge inside called Black ginger is grown by the inhabitants of Mizoram. Wild relatives of large cardamom (*Amomum subulatum*) and cinnamon are available in the forests of this region.

Ornamental Plants

The ornamental plants, which are found in NEH region, include species of Mannolia, Rhododendron, Cassia, Erythrina, Calustemom, Dacasenda, Myrica, Bauhinia, etc. About 600 species of orchids are naturalized in great proportion in this region. Wide range of variability has been found in Sikkim, Arunachal Pradesh, Meghalaya and Manipur.

Medicinal and Aromatic Plants

Medicinal plants of known use like *Rauvolfia serpentina*, *Solanum khasianum*, *Dioscorea proceri*, *Coptis teeta* are widely available in the region. Orchids as medicinal agent have been used by the tribals of NE states like Arunachal Pradesh, Nagaland, Meghalaya and Mizoram.

Germplasm Evaluation

In order to facilitate effective utilization of plant genetic resources, it is important that the plant genetic resources are evaluated for productivity including components, crop duration, resistance to biotic and abiotic stress and quality of produce.

A dwarf and late maturing polyembryonic mango cultivar Moresh bears sweet fruit with high pulp content within 2 years from planting and is free from stone weevil. The locally available species like *Malus baccata* is widely used as rootstock of apple while *Pyrus pashia* is a common rootstock of pear. The ripe fruits of *Elaegnus*, viz *E. latifolia* and *E. pyriformis* are edible and used for making refreshing drink (Pandey, 2002). The fruits of *Docynia indica* and *D. hookeriana* acidic, greenish with red tinged are eaten fresh and in pickles as well as in jelly preparation. Some of the brinjal varieties have excellent quality in having large size, soft flesh and less seeds. An important species of medicinal importance (solasodine content) is *S. khasianum*. Another species *S. torvum* is extensively used in Ayurvedic medicine system.Three tomato varieties namely Manileima, Manikhamnu and Manithoibi were released by State Variety Release Committee, Manipur and found suitable for rice-based cropping system. *L. pimpinellifolium* is also good source of resistance to late blight and tomato leaf curl virus (Seshadri and Srivastava, 2002). In chilli, a collections from Tezpur (Assam) has been found to have the highest capsaicin content recorded so far anywhere in the world.

Cho-Cho (*Sechium edule*) produce large starchy edible roots in addition to fruits. The National Bureau of Plant Genetic Resources, Regional Station, Shillong is maintaining more than 10 germplasms of Indian bean. Winged bean has excellent nutritional qualities particularly being very rich in protein (Rao and Dora, 2002). In the trials conducted by ICAR Research Complex for NEH Region, Umiam varieties of sweet potato such as S-162, S-69 and S-30 have been found to be promising. Recent studies indicate the dioscorea hybrid H-312 and H-1687 are more adapted to wide range of conditions ensuring higher tuber yield. Tree tomato is consumed as delicious chutney when raw or after roasting and peeling off the skin. The tubers of *Vigna vexillata* are rich in carbohydrates and minerals. The tubers of *Flemingia vestita* are rich in iron (2.64 mg), phosphorous (64.06mg) and contain fair amount of protein (2.99 g), calcium (19.77) and carbohydrates (27.02g) (Sarma, 2001).

Turmeric variety Megha turmeric-1 (earlier known as RCT-1) and ginger variety Nadia were found suitable for the region. Lakadong is a turmeric variety found in this region, which has high percentage of curcumin content (7.4%).

The native species of orchids having ornamental value and market potential usually belong to Aerides, Anachnantha, Arundina, Cymbidium, Dendrobium, Paphiopedillium, Phaius, Renanthera, Phycostyllus and Vanda etc. (Borthakur, 1992). The juice from cut leaves of *Cymbidium giganteum* is used by Khasis for clotting of blood while the juice of Vanda flower is used as eye drop for cure of glaycoma. There is immense scope of medicinal and aromatic plants for possible use of local germplasm in the improvement of vegetable crops for the use in horticultural industry (Borthakur, 1992).

Conservation of Diversity

Constraints

- Land tenure issues: Land tenure systems vary widely among different North-Eastern states, which are quite different from the rest of India. The complexity in land ownership and tenurial rights makes it difficult for survey, demarcation and consolidation of land. Therefore, cadastral survey and land demarcation are completely absent in the hill areas of northeast.
- Gender and equity issues in natural resources and diversity management: Unequal distribution of land resources is responsible for increasing dependence on forests by certain sections of the society leading to diversity degradation. Resolving the gender and equity issues concerning natural resource management is equally important in North-East as in the other parts of the country.
- Inter-departmental coordination: Inter-departmental coordination is essential to the sustainable management of horticultural resources in the region.
- Smuggling of timber across the international border: The illicit felling of trees and timber smuggling across the international

borders has been the most important cause of horticulture areas/forest degradation in border.

- Shifting cultivation: Unregulated shifting cultivation by the local tribal populations has been a major threat to sustainable diversity management particularly in unclassed and community forests of the region.
- Inter-state border dispute: There exist a lot of inter-state border disputes among the northeastern states. Most of these border areas are forest lands and because of boundary disputes, such lands are often declared as 'no man's land' and hence, do not come under any form of management. This leads to the degradation of diversity in such areas.
- Insurgency: The long insurgency problem in some states such as Assam and Tripura has considerable impact on diversity conservation.

Prospects

Although there are not many agencies/organizations working exclusively for diversity conservation in north-east *per se*, the activities taken up by many organizations including nongovernmental and traditional institutions, government departments and scientific institutions have direct or indirect implications for diversity conservation.

- State Government Agencies: Many state agencies are now involved in such diversity conservation activities as establishment of germplasm banks for horticultural crops.
- Research Organizations: Many state and central government research organizations including universities of the region are engaged in research, inventory and conservation of diversity in the region. Such organizations are Botanical Survey of India, Shillong, G.B. Pant Institute of Himalayan Environment and Development, (North-East Unit, Itanagar and Sikkim Unit, Panthang), Indian Council of Agricultural Research for North-Eastern Hill Region, Barapani, Shillong with campuses through out the north-east, State Forest Research Institute, Itanagar, NBPGR, Shillong, North-Eastern Hill University, Shillong, Nagaland University, Kohima, Mizoram University, Aizawl, Arunachal University, Itanagar, Tripura University, Agartala, Assam University, Silchar, Tezpur University, Tezpur, Gauhati University, Guwahati, Assam Agricultural

University, Jorhat, Regional Research Laboratory, Jorhat, Dibrugarh University, Dibrugarh.

- Non-Governmental Organizations: Many non-governmental organizations are now working for the conservation of diversity in north-east, although most of them are local and grassroots level.
- International Donor Agencies: International donor agencies in Meghalaya, Manipur and Assam, and Nagaland have been playing crucial role in conserving the diversity through their respective projects.
- International and National Policies and Conventions: All the international treaties and national policies have significant impact on the conservation of diversity in the north-east.
- Academic Institutions Including Schools and Colleges: The educational curriculum in the universities, colleges and schools has an important role to play in diversity conservation.
- Shifting Cultivators: The shifting cultivators and other traditional farming communities of northeast have played a key role in conserving the rich horticultural crops germplasm of the region. In spite of the availability of many hybrid and high yielding varieties these farmers have been cultivating the traditional varieties for generations.

Gaps

The depletion of diversity and inadequacy in actions to conserve the diversity of the region may be attributed to several factors, which range from inadequate knowledge about diversity and its components to adoption of wrong and inappropriate policies by the concerned stakeholders.

Gaps in Knowledge and Information:

- Information on urban diversity is scanty
- Species inventory in inaccessible areas of Arunachal Pradesh, Nagaland, Karbi Anglong and North Cachar hills of Assam, and parts of Mizoram and Manipur is yet to be made.
- Information on genetic diversity is extremely poor
- Information on wild ornamentals and aromatic plants is scanty.
- Gaps in vision: Most of the programmes and activities being undertaken by the state governments are shortsighted. Long-term planning based on sustainable development strategies

and integration of diversity conservation issues with development planning is the need of the hour.

- Monoculture plantations: In order to increase the revenue generation, the State Horticulture Departments pursue the policy of raising plantations of commercially important species by clearing and burning the natural diversity areas.
- Introduction of high yielding varieties/hybrids of crops: The horticulture departments are introducing various high yielding varieties/hybrids of horticultural crops. This is associated with increasing use of inorganic fertilizers and chemicals for plant protection. Such policies not only ignore the indigenous species and varieties but also have adverse effects on existing flora and fauna.
- Conversion of natural vegetation areas into cash crops: Cash crops such as tea, coffee, cardamom, ginger and turmeric are being encouraged by several agencies throughout the northeast. Many shifting cultivation rehabilitation programmes introduce these cash crops as alternatives to shifting cultivation. These plantations are obviously taken up on lands with natural vegetation (in past or present) at the cost of diversity.

Gaps in Policies and Legal Structure:

- The wrong conservation policies with focus on economically important species have been harmful to diversity. Such policies as adopted in Tripura, Mizoram, Nagaland and Meghalaya have not only decreased the species diversity in natural/ rehabilitated forests but have also resulted in accelerated soil erosion and loss of soil moisture.
- The policy of rehabilitation of *jhumias* through rubber plantation as has been done in Tripura may prove to be a disaster for other floral species in such areas.
- The policy of promoting high yielding varieties and assessment of progress and success on the basis of consumption of fertilizer and plant protection chemicals has led to ignoring the indigenous varieties. The government subsidy and credit policy is instrumental in adopting these schemes.
- Through the Public Distribution System only HYV are distributed. There is a need to include distribution of indigenous varieties too.
- The planners have not considered the role and value of diversity in preparing developmental plans. Such ignorance has been

responsible for taking no efforts to conserve and enhance diversity.

- Most of the problems are related to increase in population. The rate of population growth in the north-east is unusually high. This causes tremendous strain on the natural resources and adoption of certain policies that are not very friendly to conserve diversity. No population policy has been adopted for future planning.
- Education policy does not include teaching on diversity conservation. The school curriculum should be able to mould the young minds in favour of diversity conservation.
- There is no policy as such to create awareness among masses for diversity.

Gaps in Institutional and Human Capacity:

- The number of trained taxonomists in the region is grossly inadequate. This is one of the most important bottlenecks for completing the inventorization of diversity.

Medicinal and Aromatic Plants in Africa

Africa is the world's second largest continent after Asia, both in terms of area and population. The mainland and the island nations in the southwestern Indian Ocean and Atlantic Ocean constitute the floristically rich and diverse sub-Saharan Africa. Africa has a rich diversity of flora with an estimated 68,000 plant species of which about 35,000 are endemic to the continent. Madagascar has the highest rate of species endemism. About 82% of a total of 10,000 to 12,000 species are endemic to the country. The lowland rainforests of Cameroon, Côte d'Ivoire, Gabon, Liberia and Nigeria; the montane forests of eastern Congo, Rwanda and western Uganda; the coastal areas of Kenya and Tanzania; and the arid lands of Ethiopia, Namibia and Somalia are known as the major centres of species endemism.

Medicinal plants, and the drugs derived from them, are the most important and readily available source of health-care remedies to rural people in Africa. In eastern Africa, many biological resources are used for obtaining pharmaceuticals that have a high national and international economic value. Western Africa is the home to one of the world's largest rainforests constituting many medicinal plant species of commercial importance. There are about 3,000 medicinal plant species in southern Africa of which 10% have found common and widespread uses in traditional healthcare systems. Northern

Africa has about 10,000 plant species of which around 70% are known to be valuable as food and medicines among other uses. Over 10% of the region's floristic diversity has potential for commercial exploitation and half of them are underutilized.

About 80% of the African population relies on traditional medicine for healthcare needs. Some people use traditional medicine only, while others combine it with conventional drugs. The use of medicinal plants by local people accounts for 70% or more of basic health-care treatments in Africa. Traditional medical practitioners (TMPs) are a crucial component of the health-care delivery system. Medicinal plants are used mainly in local, traditional medicine rather than exported to foreign markets. Demand is increasing and often exceeds supply.

The health budget for Africa is largely spent on conventional medicine and is not sufficient to meet requirements.

Public expenditure on health is about 2% of gross domestic product (GDP) and access to health services and sanitation stands at 64 and 49% respectively. In addition, the ratio of modern medical doctors and facilities to population is inadequate. The State cannot afford to provide costly modern medicines for everyone. As a result, most of the rural population relies on TMPs.

Africa accounts for only 5% of global pharmaceutical trade. However, for crude drugs it is next to Asia for the export of medicinal and aromatic plants.

Medicinal plants are a common item at urban markets. Most medicinal plant material is exported to Europe and the rest to the USA and Asia. In 1996 about 26,500 tonnes of medicinal and aromatic plant material were exported to Europe.

Most medicinal plant material is sourced from wild collection and some from cultivation. Over-exploitation and habitat destruction of commercially important medicinal plant resources threatens their existence in the wild. The World Bank, the International Development Research Centre (IDRC), and the Danish Agency for Development Assistance (DANIDA), among many other international agencies, are making considerable efforts to conserve African biodiversity. The Missouri Botanical Garden, St. Louis, the USA is collecting specimen samples of tropical flora from Cameroon, Congo, Equatorial Guinea, Gabon, Madagascar and Tanzania, and samples of temperate flora from South Africa for anticancer and anti HIV/AIDS screening programmes of the United States National Cancer Institute. The

Royal Botanic Garden, Kew, London, the UK, has a similar project in Africa.

Many African countries have made efforts to formulate policies on traditional medicine, recognized the role of TMPs and established departments of traditional pharmacopoeia. Despite these efforts, many constraints and problems hinder the development of traditional medicine in Africa. Herbal medicines have not been officially recognized in most countries and their regulation and registration have not been established. In countries with apparent recognition, appropriate budgeting to facilitate the functioning of traditional medicine authorities is inadequate or totally lacking. The traditional medicine community operates outside the framework of national legislation for collection and trade in wild species. Many African countries do not have procedures to register medicinal plant preparations although these are widely used by the majority of population.

Regulation, if any, is very stringent, requiring the same standards as modern medicines. In brief, it can be concluded that the medicinal plant sector in Africa is highly unorganized.

The increasing interest on medicinal and aromatic plants, the global resurgence of plant-based medicines and their potential to provide viable alternatives to conventional drugs prompted the International Centre for Science and High Technology (ICS-UNIDO) to prepare a document on the status of medicinal and aromatic plants in Africa. The document presents in-depth information on medicinal plant resources including R&D activities, and the trade and marketing situation in African countries.

The intention is to help policy-makers, the scientific community and entrepreneurs to frame effective policies, formulate projects to augment the R&D activities and create an environment conducive to the growth of plant-based medicine. It will help African countries to improve the health of their populations, using their own resources and a system of medicine which is less expensive than modern medicine and already accepted by African society.

9

Climate Variability and Crop Production in Arid

Western Rajasthan

The hot arid region of India extends over an area of 317,090 km2, covering seven administrative states of India viz., Rajasthan, Gujarat, Haryana, Punjab, Maharashtra, Karnataka and Andhra Pradesh. Two states, Rajasthan and Gujarat, account for 81.5% of the total hot arid areas in India (Krishnan, 1968a, b). There are different methods of characterizing aridity of an area. In order to take up development work that would provide the largely agriculture-dependent inhabitants the capabilities to withstand impact of drought and other natural hazards in the drylands, including arid areas, Government of India followed the concept of demarcating arid and semi-arid areas on the basis of tehsil-wise water balance to determine moisture index. If the moisture index was below -66.6%, the area was considered arid and if it was between -33.3 to -66.6% it came under semi-arid zone. Based on this demarcation areas were identified for Desert Development Programme (in arid areas) and Drought Prone Area Programme (in semi-arid areas).

Climate and its Variability

Arid Rajasthan is characterized by limited seasonal precipitation with erratic distribution, high atmospheric temperature that has large diurnal and seasonal variation, strong insolation and persisting wind regime. Consequently, there are high crop water requirements. The weather conditions remain too dry, even in normal years, for most part of the year and are inhospitable for successful crop growth.

Rainfall and its Distribution

Normal dates of sowing rain (>25 mm week-1) in western Rajasthan are between 1st and 15th July, but the monsoon sometimes sets in as late as 3rd week of July in the eastern part and 1st week of August in the western part. Sowing rains have also been found to occur at times as early as 1st week of June in the eastern part and 2nd week of June in the western part. Such early or late commencement of sowing rains lead to large variability in yearto- year crop productivity. The assured crop growing period in western Rajasthan varies from 25 to 90 days under shallow soils and 35 to 105 days under deep soils (Rao *et al.*, 1994).

The mean annual rainfall in the region varies from 185 mm at Jaisalmer to more than 467 mm at Sikar. About 80-90% of the annual rainfall is received during the southwest monsoon when westsouthwesterly moving depressions are the main sources for *kharif* crops. The depressions originating from the Arabian Sea during May or early June move in east-northeasterly direction and also cause rainfall but such events occur with a lesser frequency of once in 2 or 3 years. The western disturbances during winter also cause precipitation, especially in the northern districts, which favour the *rabi* crops. The high inter-annual variability in rainfall is reflected in its high degree of coefficient of variation, ranging from 36% in the eastern part to 65% in the western part. This variability in annual rainfall is the most important factor, influencing the crop and pasture yields in the region.

Solar Radiation and Duration of Sunshine

Solar radiation is generally high throughout the year in western Rajasthan. During winter, it varies between 15.12 and 17.71 MJ m-2 day-1 and in summer months the values range from 22.79 to 26.50 MJ m-2 day-1, with a mean of 22 MJ m-2 day-1. The daily duration of bright sunshine hours in the area remains above 10 h day-1 in May and reduces to 6.6 h day-1 in July and August. In winter it is above 8.8 h day-1.

Air and Soil Temperatures

The region experiences extreme air and soil temperatures, which considerably increase the demand for water requirements by vegetation/crops. During winter, mean maximum air temperature varies from 22.4°C to 29.0°C and minimum temperatures between 4.1°C and 14.3°C. Air temperatures sharply increase from April onwards and stands highest during May till pre-monsoon showers sets in the area. Summer air temperatures vary between 31.2°C and 42.0°C with peak values as high as 50.0°C. Temperatures fall during the monsoon period (June-September), but rise by about 3.0°C to 5.0°C after the recession of the monsoon and again start falling up to -5.7°C from December onwards due to winter conditions. Soil temperatures follow the diurnal and annual cycles of air temperatures, and shoots up to 62.0°C during May and June, when in general, it is higher than air temperatures by 10.0°C (Ramakrishna *et al.*, 1990).

Relative Humidity

Relative humidity in the region is often less than 30% during summer months, but gradually increases to 80% by monsoon and then decreases from October onwards, following the withdrawal of the monsoon. Low humidity, combined with strong wind regime, leads to advection, a phenomenon that causes evaporation loss more than the energy actually available through solar radiation.

Wind Regime

The winds over the region blow from southwest during the monsoon and from northeast during the winter. Winds during the winter are low. From April onwards a strong wind regime builds up along with the increasing temperature, and reaches 15 to 18 km h-1. Peak winds occasionally reach as high as 60-80 km h-1 during severe dust storms. Strong wind regime during May and June cause wind erosion, depleting soil fertility.

Evapotranspiration Rates

The extreme climatic conditions in the region result in high evapotranspiration rates. The potential evapotranspiration rates are 1.3 to 3.5 mm day-1 during winter months of December and January. By May, the values rise to 3.7 to 9.5 mm day-1. During monsoon period the evapotranspiration rates are 5.2 to 10.6 mm day-1, but decrease from October onwards till December. The annual potential evapotranspiration ranges from 1400 mm in the eastern part to more than 2000 mm year-1 in the western part of arid western Rajasthan.

Weather Extremes

Frequency and Nature of Drought

The region is frequently prone to severe drought due to low and erratic rainfall. The frequency of agricultural drought at four locations of arid Rajasthan, as per Ramana Rao *et al.* (1981). Out of 108 years

(1901-2008), the region experienced agricultural drought in one part or the other in 52 to 62 years, which suggest drought occurs in the region once in three years to alternate year.

Jaisalmer district is most prone to drought. During 1901-2008, the agricultural drought occurred here in 70% of the years, out of which 43% of the years experienced severe drought and 27% years moderate. Barmer district experienced severe drought in 30% years and moderate drought in 20% years. Bikaner experienced severe agricultural drought in 23% years and moderate in 25% years, whereas Jodhpur district experienced severe drought in 17% years and moderate drought in 28% years. Often, drought persists continuously for 2 to 6 years. Prolonged drought was experienced in the region during the decades 1901-10, 1911-20, 1932-40, 1961-70, 1981-90 and in 2001-2006, with 1918 and 2002 standing out as the worst drought years during the recorded period. The moderate to severe drought frequency varies from 1 out of 5 years in Churu district to 2 to 3 out of 5 years in Barmer and Jaisalmer districts.

Probability of occurrence of early, mid and late season drought The weekly rainfall pattern at Jodhpur during 1971 to 2008 indicated that the rainfall distribution was fairly good in 17 years, whereas in the remaining 21 years, the region experienced drought at one stage or the other. Further classification of these 21 years' drought, based on weekly rainfall distribution, showed that the probability of early season drought was 26% (9 years), mid-season drought in 9% (3 years) and late season drought in 26% (9 years). Among the three categories of drought, the impact on grain yield of pearl millet and pulses was felt more under mid-season and late-season drought. If the drought is severe at early season of the crop and reduction in rainfall in subsequent crop stages, the impact of such severe drought is felt in the region on food, fodder and drinking water.

Impact of drought on livestock population

The region supports a large livestock population, out-numbering humans by about 4 times in Jaisalmer district, 3 times in Barmer, 2 times in Bikaner and 1.5 times in Banaskantha, Jalor and Jodhpur districts, as against only one-half in the rest of the country. Ahuja (1994) showed that the supply of forage was 37% shorter than the demand at the current level of livestock population. During severe drought years, the normal forage productivity was declined by 82% in rainfall zone of <200 mm, 74% in 200 to 300 mm rainfall zone and 42% in 300 to 500 mm rainfall zone (Shankarnarayan and Singh.

1990). Consequently, during drought years, the livestock population decreases in the Indian arid zone due to scarcity of fodder and water. Sheep, goat, buffalo and cows were the worst affected during drought years. The consecutive drought during 1985, 1986 and 1987 resulted in decrease of livestock population (26% in cattle, 7% in sheep; Source: DAH, 2005).

Flood Scenario

Floods in this arid region are rare as compared to drought. The decadal-wise frequency of the flood-causing rainfall (called flood years), however, indicated that it occurs with a frequency of 2 to 3 out of 10 years. The decades 1911-1920, 1941-1950 and 1981-1990 were notable in this respect. Broadly such rainfall has been recorded in alternate decades, e.g., in 1970-1979 and in 1990-1999. Widespread flood-causing rainfall was witnessed during 1908, 1917, 1944, 1975, 1983 and 1997.

During

1917, the average rainfall in the region was exceeded by +161%. Such widespread excess rainfall was not recorded in subsequent years (Rao *et al.*, 2006).

Cold and Heat Wave Conditions

Extreme temperatures are frequently encountered in arid Rajasthan, which cause great threats to the growth and productivity of agricultural crops in both *kharif* and *rabi* seasons. The cold or heat wave conditions many times follow in a sequence, affecting crops at various stages of growth.

Often air temperature falls below -4.4°C in winter, resulting in frost injuries in *rabi* crops like mustard, castor, cumin, gram, barley, taramira and vegetables. In association with passing western disturbances, severe cold wave conditions prevail during December and January. The districts of Churu, Jhunjhunun, Ganganagar, Hanumangarh, Sikar, Jaisalmer and Bikaner are more affected by intense cold wave conditions.

Climate Change in Arid Rajasthan

The impact of projected climate change by the end of 21st Century (IPCC, 2007) is more likely in arid region than in the semi-arid or sub-humid regions of India. The arid phase of northwest India has a history of about 3000 years (Pant and Maliekel, 1987). In the northwest India, covering Punjab, Haryana, west Rajasthan and west

Madhya Pradesh, there was a marginal increase in the rainfall by 141 mm and fall in air temperature by - 0.52°C in the past 100 years (Pant and Hingane 1988; Rao and Miyazaki, 1997). The secular trends in annual rainfall in the arid western Rajasthan showed no significant change during 1901 to 2008.

To know the impact of introduction of large-scale irrigation through *Indira Gandhi Nahar Parjyojana* (IGNP) canal into the region, past hundred years of annual rainfall and air temperature data were analyzed from selected locations along the tracts of the canal (Rao, 1996, 2005). Long-term annual rainfall at Ganganagar showed an overall increasing trend at the rate of 1.18 mm year-1. Even during drought years, Ganganagar area received higher rainfall than the other arid areas, attributed to the irrigation imposed for longer periods that supplied the moisture to rainfall systems and also indirectly through energy balance from increased vegetation. However, increase in rainfall was not observed in adjoining Bikaner and Jaisalmer districts, where the irrigation was introduced since past 1-3 decades.

Strategies to Cope up Weather Extremes and Climate Change

Early Warning of Drought and Crop/Forage Yields

Simulation of crop production from arid regions had several difficulties due to temporal and spatial variability in rainfall, high evapotranspiration rates and heterogeneity in land use or crop acreage. Because of our limited knowledge in understanding the assimilation of carbon by various arid zone crops, by and large we have to depend on regression models or soil moisture simulation models for explaining the crop/forage yield.

For drought and crop surveillance over the region, computer models like the SPAW (Soil-Plant-Air-Water) and CERES-millet have been tested and modified at CAZRI. The SPAW model was used for assessment of soil moisture and crop stress conditions under pearl millet, which could explain 89% of pearl millet yield in Jodhpur district.

The relationship between pearl millet grain yield (kg/ha) and water stress index (WSI) of SPAW model was Y = -45.38 WSI + 526.18 (r=0.9427) (Rao and Saxton, 1995; Rao *et al.*, 2000). Of the various crop growth simulation models developed by the International Benchmark Sites Network for Agrotechnology Transfer (IBSNAT), using the DSSAT (Decision Support System for Agrotechnology

Transfer), the CERES-millet model was used for simulation of pearl millet (Ramakrishna *et al.*, 1994), which predicted grain yield and other parameters closer to the observed yield.

RANGETEK model was used for prediction of forage yield of promising arid grasses like *Lasiurus sindicus, Cenchrus ciliarias, C. setigerus* (Singh *et al.*, 1996a, b). The model was used for real-time simulation of daily soil water content, soil evaporation and plant transpiration to forecast forage production in terms of a ratio of actual to potential transpiration (yield index). The deviations between recorded and estimated forage yield using the model were from 8 to -12% (Singh *et al.*, 1996a). The studies on *C. ciliaris* revealed that the predicted yields using the RANGETEK model were between -10% and +15% of actual forage records (Singh *et al.*, 1996b). These 1-dimension model combined with a knowledge based geo-spatial Decision Support System (DSS) could be an effective tool for drought early warning and management by farmers, experts and general end-users.

Evapotranspiration Rates and Water-use Efficiency of Crops

Evapotranspiration requirements of crops were measured in gravimetric lysimeters, keeping the crop under unstressed conditions (100%PET rates of irrigation). The crop coefficients at different growth stages, water-use efficiency (kg/ha/mm) were computed for planning scarce water resources of arid area for increasing the productivity per unit of water.

Rainwater Harvesting for Coping with Drought

Rainwater harvesting is an old practice in arid Rajasthan. Both *insitu* as well as *ex-situ* techniques can be utilized to harvest runoff rainwater for productive uses like drinking and supplementary irrigation purposes. It is estimated that nearly 300 mcm of water for human consumption, 350 mcm for livestock and 6800 mcm for irrigation is required by 2010 (Venkateswarlu *et al.*, 1990).

Crop Contingency Plan for Delayed Monsoon Condition

CAZRI (Narain *et al.*, 2006) prepared contingency plans for delayed monsoon. Delay in monsoon reduces the water availability period and need strategies to grow crops, which match and yield better under such conditions.

Trends in Arid Zone Research in India

Such contingency plans were found useful to obtain better economic returns compared to traditional systems.

Use of weather based Agro-Advisory Service

Weather based agro-advisory bulletins to farmers provide information on a real-time basis so that they can avail advantage of climatic information to minimize crop losses caused by unfavourable weather. Since 1998 CAZRI is issuing bi-weekly agro-advisory bulletins based on Medium Range Weather Forecasts prepared by National Centre for Medium Range Weather Forecast (NCMRWF). The information on the type of weather situations likely to be encountered and the method to be adopted for efficient management of inputs, is provided to the farmers in advance for minimizing weather related risks.

To measure the economic impact of agro-advisory services (AAS), a three-year survey in Manai, Narva and Palri Mangalia villages of Jodhpur district was conducted during 2003 to 2006. The economic analysis on benefit-cost ratio for *kharif* and *rabi* crops indicated that for pearl millet the benefit-cost (B:C) ratio was 2.15 to 3.17 for AAS farmers, whereas for non- AAS farmers it was 0.73 to 2.80. In case of mustard, farmers were much benefited as the B:C ratio for AAS was 1.92 to 2.25 against 1.64 to 2.13 for non-AAS farmers (Rao, 2008).

Land Use of Arid Lands in India

Land use is the human modification of natural environment of wilderness into built environment. Land cover is the observed bio-physical cover on the earth's surface while land use is the arrangement, activities, and inputs people undertake in a certain land cover type to produce, change or maintain it (FAO, 2005). India has about 18% of world's human and 15% of livestock population to be supported with only 2.4% of the world's geographical area and 1.5% of forest and pasture land. Per capita availability of land in the country has reduced from 0.4 ha in 1950-51 to 0.14 ha in 2000- 01. With the prevailing trend it may be further reduced to 0.077 ha by 2020.

In the hot arid ecosystem of India low and erratic rainfall, extremes of temperature, high evaporation loss, saline and meagre groundwater, absence of perennial streams, and dune-covered and rocky/gravelly terrain, are the major factor influencing the land uses. Agriculture, which is the most dominant land use, is mainly rainfed and subjected to high risk and uncertainty. The cold arid Ladakh region of Jammu and Kashmir state is situated in the high Himalayan Mountain, where arable land is very limited and the length of growing period (LGP) is short. Human activities are very limited and so are the livelihood sources.

Before the 1960s land use studies were primarily based on revenue and land records data. Thereafter the use of topographical maps and aerial photographs made it possible to show the spatial distribution of different land use systems. After the 1970s satellite remote sensing and GIS virtually revolutionized the research on land use mapping and monitoring of changes over time and space. Provides a short review of the major findings on land use studies on arid ecosystem of India.

Land Use: A Historical Perspective

History of land use and land classification system in the arid ecosystem is reported by Roychaudhury (1966, 1970). Evolution of land use in north-west arid zone of India has been reported by Whyte (1960), Bharadwaj (1965) and Randhawa (1980). According to these studies land use in the region evolved through the Neolithic Period when man had taken up sedentary life and had domesticated the animals.

They first settled along the rivers and near water points, and later on shifted to other marginal areas. James Todd's Annals and Antiquities of Rajasthan (1929; reprinted 1978) and Erskine's Western Rajputana States Residency and Bikaner Agency Gazetteer (1909) highlight the land use system that prevailed during British Period but no systematic account is given. Classical treatise of Nainsi Mohto (Singh, 1969; Jain, 1974) on *Marwar-ra Pargana-ri Vigat* provides an account of cultivated area and crops being produced in the 'Khalsa' villages which were directly administered by the rulers. No account was available for villages under the *Jagirdars.*

After the settlement of land holdings, abolition of *Jamindari* system and reorganization of states, the use of each land holding, including the area under different crops as per cadastral map of a village began to be recorded in a register called the *Girdawari*, and it continues till today. Kanitkar (1952) estimated that out of 88597 miles2 area in Rajputana desert only 20694 miles2 was available for cultivation, while net sown area was 8124 miles2. Irrigated area was nearly 1486 miles2. Banerji (1952) while defining the role of vegetation in desert control estimated 320 miles2 of *'Khalsa'* area under forest department. Total *khalsa* area was 5,545 miles2.

Thus forest covered only 5.8% area. Culturable waste covered 523 miles2. Nearly 31,000 miles2 was with the *Jagirdars.*

Land Use Classification System

The Land Records Office at tehsil or taluka level keeps land use record of each village under their jurisdiction using a 44-column proforma. This classification system categorizes fallow lands into three sub-classes under individual and government holdings; area irrigated by different sources and subclasses of rainfed, double cropped and net sown areas. Based on this format the State-level Directorate of Economics and Statistics compiles and publishes annually the land use data by adopting a 16-point format and 12 land use classes. Irrigated area is not included in this classification system, but is given separately under irrigation. At national level the land use data is published by the Directorate of Economics and Statistics, Department of Agriculture and Cooperation, under a nine-fold classification system. The district Census Handbook published by the Directorate of Census Operations also includes land use data in five classes, viz., forest, irrigated area, rainfed area, culturable waste and permanent pastures, and area not available for cultivation. The spatial pattern of the land uses was partly available from the large-scale topographical sheets prepared by Survey of India.

Systematic mapping of land uses in the arid zone began after the establishment of Central Arid Zone Research Institute in 1959, especially as a component of integrated basic resources survey and mapping. Taking into consideration various land use classification systems available then, the typical problems of arid zone and the feasibility of mapping, A.K. Sen evolved a land use classification system for the Indian arid zone, which is summarized in Sen (1978, 1982). Sen (1972a, I977a, b) also developed aerial photo interpretation techniques to classify and map different land use categories. The classification contains 15 land use categories. Cultivated lands were classified into seven categories based on the intensity of cultivation to avoid repetition of land use survey every year, since the annual extent of cropped area varies and depends on the intensity, distribution and amount of rainfall. Wastelands were classified into four categories viz., sandy, saline, rocky/stony and rocky and gravelly waste, to which were added subsequently the gullied waste, mud waste and waterlogged areas. It formed a basis for the future nation-wide land use/land cover mapping from satellite data (Gautam and Narayan, 1988; NRSA, 1989) that identifies six land use/land cover categories under Level-I and 24 categories under Level-II, as well as the wasteland mapping units (NRSA, 1986, 1991). The satellite-based classification system was further refined during the Integrated Mission for Sustainable Development (IMSD) project of ISRO, when classification up to Level-III was possible (i.e., kharif and rabi croplands; dense and open forest/grassland cover; NRSA, 1995a; Balak Ram, 2002). The classification system has further evolved in recent years with the availability of higherresolution satellite images.

Present Land Use

As per land use statistics of 2005-06 out of 42.5 mha area of hot arid ecosystem, cultivated lands constitute 63.67%. This includes 51.69% net sown area, 6.26% current fallow and 5.72% other fallow land. Net and gross irrigated areas are 17.26% and 26.75%, respectively. Double cropped area is 14.09%. Among the other uses, forestland occupies 5.72%, pasture 3.21%, land put to non-agricultural uses 5.3%, barren and uncultivable land 9.10% and miscellaneous tree crops and groves 0.18%. Land use statistics of the hot arid zones of Rajasthan, Gujarat, Punjab, Haryana, Andhra Pradesh and Karnataka as per revenue. Satellite-based mapping has revealed that the region has 9.12 mha or 21.56% area under wastelands (NRSA, 2005) of different categories. Arid western Rajasthan has its 29.4% area under wastelands, Andhra Pradesh 22.27%, Gujarat 13.17%, Karnataka 11.22%, Haryana 3.58% and Punjab 4.07%. Sandy wastes are most dominant in Rajasthan, barren rocky/stony waste and salt affected lands in Gujarat, and degraded forest and land with/without scrub in Andhra Pradesh and Karnataka. Pearl millet, cluster bean, moth, mustard and wheat are the important crops in arid Rajasthan, cotton, wheat, rice and mustard in Haryana and Punjab, groundnut, cotton and castor in Gujarat and green gram, red gram, groundnut and sunflower in Andhra Pradesh and Karnataka.

In the cold arid region, out of 9.67 mha total area (in Leh and Kargil districts) the reporting area is only 64,626 ha. Rest of the area is covered by snow and hill ranges. Out of the total reporting area cultivated lands constitute 30.92%, which includes 29.67% net sown area, 0.53% current fallow and 0.72% other fallow land. Forest occupies 0.1% area, land put to non-agricultural uses 6.32%, barren and uncultivable land 45.03%, pasture 1.69%, miscellaneous tree crops and groves 2.49% and culturable waste 12.36%. Wasteland statistics of NRSA (2005c) revealed 5.06 mha (88.28% of the total geographical area) under wastelands in the two districts. This includes 52.6% barren rocky/stony waste, 12.05% steep sloping area, 30.15% snow covered area, 2.68% sands and 0.18% degraded pastures. Ragi, wheat, barley, millet and apricot are important crop of this region.

Land Use Changes

Between 1982-83 and 2005-06 net irrigated and double cropped area in hot arid region has increased by 128% and 70%, respectively.

Net sown area has increased by 2.4% only, current fallow by 1.7% and other fallow land by 12.1%. More impetus has been given to irrigated cash crops. The area under forest has increased by 4.6% and land put to non-agricultural uses by 16.6%. On the other hand, the area under culturable waste has declined by 18.6%, pasture by 7.7%, miscellaneous tree crops and groves by 36.8% and barren and uncultivable land by 2.2%.

Arid western Rajasthan has recorded substantial increase of 42% in net sown area, while forest has increased by 67.4%. Area under culturable waste has declined by 26.3%. This shows that though marginal lands have been brought under cultivation, the intensity of cultivation, particularly in rainfed lands, has not increased to that extent. This has increased the fallow lands. Environmentally sensitive common lands have become more victims of degradation by nature and man. Uneconomic use and over-exploitation of natural resources have caused problems like secondary salinization, waterlogging, mined spoil areas, degradation of groundwater, intensification of sand movement, blockade of natural drainage and spread of industrial effluents. Silting and pollution of irrigation tanks/ponds, degradation of mangroves and reduction in biomass are the other impacts.

In the cold arid region, horizontal and vertical development of agricultural is not up to the mark. Between 1982-83 and 2005-06 the net sown area has increased by 5.1%, other fallow land by 151.1%, land put to non-agricultural uses by 37.23%, and double cropped area by 86.3%. On the other hand the area under culturable waste has declined by 19.16%, miscellaneous tree crops and groves by 45.9% and current fallow by 52.77%. As per forest department there is 13166 km^2 forestland and 13130 km^2 area under wildlife conservation.

In western Rajasthan, Directorate of Agriculture, Government of Rajasthan occasionally brings out district-level trends in land use (e.g., from 1956-57 to 1990-91; Anon., 1992b). Based on field mapping, Balak Ram (1988) provided a quantitative account of land use changes in Siwana region between 1960-61 and 1985-86 at village, tehsil and regional level. Since then, several studies have been carried out on land use change mapping. Notably, Balak Ram and Rastogi (2004) mapped the land use changes in Kachchh district, while such mapping has recently been completed in Jaisalmer district (Balak Ram and Chauhan, 2009). Balak Ram and Gupta (2003) assessed the impact of industrial effluents on land use changes along the Jojri, the Bandi and the Luni rivers. Cincotta and Pangare (1991) discussed agricultural land use changes and pastoralism in Gujarat, while Chaurasia *et al.* (1996) discussed land use change in Punjab. Nautiyal *et al.* (2005) described ecological impacts of land use intensification in the Himalayas. Analyses of the earlier trends are summarized in Balak Ram and Gheesa Lal (1997, 1998) and Balak Ram (2003).

Issues and Challenges

Important issue and challenges of hot arid region are risky and traditionally subsistence rainfed farming with low productivity, degraded grazing and forest land, traditional and low productive animal husbandry, mined spoil areas, over-exploitation and injudicious use of ground water, land degradation and drought. In cold arid region the important issues are soil and water erosion, traditional use and non-managed pasture and forest land, subsistence and segregated farming system, poor system of animal husbandry and livestock production, neglected human and cultural value, under-utilized and ill-managed rain and river water. These are being constantly addressed by different government and non-government programmes through numerous institutions of research and development, but the vastness of the country, vagaries of climate and very high population pressure make the impacts manifest either slowly to, or are reversed fast.

Variability in Arid Soil

Characteristics Arid region in India is spread over in 38.7 million hectare area. Out of the total, 31.7 m ha lies in hot and remaining 7 m ha lies in cold region. The hot arid region occupies major part of northwestern India (28.7 m ha) and remaining 3.13 m ha is in southern India. The northwestern arid region occurs between 22030' and 32005'N latitude and 68005' to 75045' E, covering western part of Rajasthan, northwestern Gujarat and south-western parts of Haryana and Punjab.

In southern India it occupies parts of Andhra Pradesh, Karnataka and Maharashtra. About 62% area of arid region falls in western Rajasthan followed by 20% in Gujarat and 7% in Haryana and Punjab. Andhra Pradesh, Karnataka and Maharashtra together constitute about 11% area of arid region. The boundary between arid and semi arid region in Rajasthan cuts across the Jalore and Pali districts, goes roughly along the boundaries of Nagaur and Ajmer. The arid zone extends to Mahendragarh and Hissar districts of Haryana; Sangrur, Bhatinda and Firozepur districts of Punjab; Jamnagar, Kachchh, Banaskantha, Mehsana, Surendra Nagar and Rajkot districts of Gujarat.

The arid districts of Haryana and Punjab exactly border to north of western Rajasthan, while districts of Gujarat touch the southern boundary. Thus arid zone in India forms a contiguous tract, keeping western Rajasthan in the centre. In the arid regions of Rajasthan, Haryana, Punjab and eastern Gujarat, aeolian and alluvium are the major formations. These aeolian and alluvial parent materials are the quaternary formations. During that period this region witnessed wide spread alluvial sedimentation. However, the overall environment was semi arid (Dhir, 1989; Dhir *et al.*, 1994; Gupta *et al.,* 2000). This alluvium was reworked by the subsequent activities. Since the last glacial period the region has experienced two major periods of aridity separated by a long period of amelioration of climate during which natural vegetation reached its maximum extent (Singh, 1977). The Dune forming aeolian activities began at least 200K years ago and terminated around 13,000 years B.P. (Singhivi *et al.*, 1983). The vast alluvial plain of Haryana and Punjab was formed out of sediments deposited by the rivers flowing from the Shivaliks. During arid phase this alluvium was reworked by aeolian activity and resulted in the formation of sand dunes and sand plains (Sidhu and Sehgal, 1978).

The cold arid region occupies 7.0 m ha area covering Ladakh, Kargil districts in Jammu and Kashmir and Lahul Spiti district in Himachal Pradesh. The region is characterized with mild summer and annual rainfall of less than 150 mm. The soils occur on summits and ridge tops, mountain and valley glacier, glacio-fluvial valleys and fluvial valleys of greater Himalayas. Soils are skeletal and calcareous with alkaline in reaction and low to medium in organic matter content (Pratap Narain, 2008).

The Soils

. The soils of this tract have been mapped in Entisols, Aridisols and Alfisols soil orders. The Entisols cover maximum 17134.26 thousand hectares (54.21%), followed by Aridisols and Alfisols. The area of latter two is 14254.32 and 213.10 thousand hectares, comprising 45.1 and 0.67% of hot arid India. Haplocalcids, Haplocambids, Haplosalids, Petrocalcids and Haplogypsids constitute Aridisols at the great group level, while Torripsamments, Torriorthents and Torrifluents are the part and parcel of Entisols. The Natrargids, Pleargids and Haplargids are the great groups mapped in Alfisols. The spatial distribution of these great groups. Haplocambids dominantly occur in the arid part of Haryana, Gujarat and Punjab, covering around 77, 42 and 39.6% of delineated area in their respective state. Torripsamments dominantly mapped in 47.1% area of western Rajasthan, while the share of these soils to the marked area is around 43,14.3 and 9.7% in Punjab, Haryana and Gujarat, respectively.

Haplocalcids are the second dominant soils of Gujarat after Haplocambids with 11.1% of allocated area under arid land. Petrocalcds and Haplocalcids together constitute 8.4% area of western Rajasthan. The contribution of Haplosalids to the delineated arid land is higher in Gujarat than Rajasthan. Torriorthents, the shallow skeletal soils associated with hills are mapped in 7.8 and 4.3% area of arid part of Rajasthan and Gujarat, respectively. Torrifluents fine textured deep soils along the course of river constitute around 2% area of arid zone both in Rajasthan and Gujarat. Haplogypsids and Haploargids represent relict soils in Rajasthan and Gujarat, while Natrargids and Pleargids are the paleosols mapable only in the latter (Shyampura *et al.,* 2002).

The soils on summits and ridge tops of greater Himalayas are shallow to medium deep, excessively drained, sandy-skeletal to loamy skeletal, highly calcareous and severely eroded with low organic carbon and available water capacity (Sidhu *et al.*, 2007). While, the soils of Mountain and valley glacier are similar in characteristics except they are covered with snow for a longer period. These soils were classified as Lithic Cryorthents and Typic Cryorthents. The glacio-fluvial valleys and fluvial valleys soils of greater Himalayas are medium deep to deep, well to excessively drained, loamy skeletal to coarse loamy and fine loamy soils with slightly acidic to neutral with medium to high contents of organic carbon. Some soils are highly calcareous and slightly to moderately alkaline have been classified as Typic Udorthents/Udifluvents, and Dystric Eutrudepts great group, respectively (Sidhu *et al.*, 1997).

Physical Properties

Soils under Entisols are generally found in those parts of arid regions where high aeolian or fluvial activities are observed. The average soil depth of this type of soil is 118 cm. Average bulk density of these soils is 1.58 Mg m- 3. Surface horizon is rich in sand content in comparison to subsurface horizons. Soil texture is sandy with average sand content of 89%. In the extreme arid part at Jaisalmer, Bikaner, and Barmer sand content is > 90% for most of the places. Soil colour is mainly yellowish brown. The hue of the soil is 10YR, value ranges from 5-6, and chroma ranges from 3-4. Surface as well as subsurface soils are single fine grain in structure. Soils are mainly loose, non-sticky and non plastic in consistency. Most of the soils are well drained and fall under very rapid (>25.4 cm hr-1) class of O'Neals permeability class. Water retention at 1/3 bar (field capacity) on an average is 6.69% (0.07 cm-3 cm-3 or 70 mm ha-1 of soil depth). Similarly, the water retention at 15 bar (permanent wilting point) is also very low (2.62%) In contrast to Entisols, Aridisols are dominant in those parts of arid regions where aeolian activity is comparatively less. The average soil depth of this type of soil is 100 cm with well demarcated horizons. Concretions of calcite below the soil profile are a common feature of these soils. Average bulk density of these soils is 1.43 Mg m-3. Average sand content of Aridisols is lesser than Entisols and is around 70%. Soil colour is mainly yellowish brown to dark brown or pale brown. The hue of the soil is 10YR, value ranges from 5-6, and chroma ranges from 4-5. Surface as well as subsurface soils are single fine grain in structure. At few places, fine weak subangular blocky structure is found at subsurface soil layers. Soils are mainly loose, non-sticky and non plastic in consistency. At few places, presence of high clay and organic carbon at subsurface layers make the soil massive, friable, and slightly sticky. Most of the soils are well drained and fall under either rapid (12.5-25.4 cm hr-1) or very rapid (>25.4 cm hr-1) class of O'Neals permeability class. Water retention at field capacity (FC) and permanent wilting point (PWP) in Aridisols are comparatively higher than Entisols and are 13% and 5%, respectively. Soils under Alfisols are found in few places of arid regions of India. In comparison to Entisols or Aridisols, these soils are comparatively rich in clay content (36%). Organic carbon content of Alfisols is also higher (3.7 g kg-1) than Entisols (1.5 g kg-1) and Aridisols (1 g kg-1). Average permeability of these soils are moderately slow. Due to presence of high amount of clay, water retention at both FC and PWP are high (23% and 10%, respectively).

Nutrient Status of Arid Region Soils

Characteristically these soils are very low in organic matter/ humus and most of the nutrients reserve is present in un-weathered mineral forms. These soils have low clay and silt, and therefore nutrient adsorption and retention by these soils are very low. Soils are generally alkaline in nature and high in soluble salts and calcium contents. Organic matter and Nitrogen Arid zone soils are low in organic matter because of low vegetation cover, high temperature and coarse texture. Organic carbon content in soils below 300 mm rainfall zone ranges between 0.05-.2% in coarse textured soils 0.2-0.3% in medium textured and 0.3-0.4% in fine textured soils. Dhir (1977) has shown that even in low range of organic carbon (>0.5%) increase in clay content is associated with rise in organic carbon. With increase in clay content there is increase in organic carbon content. Joshi (1990a) has reported that bulk of organic carbon (67-89%) is present in non humic form followed by fulvic and humic form and mean value of HA-carbon/FA carbon increased with increase in silt+clay. Aggarwal and Lahiri (1977), Aggarwal *et al.*, (1993) reported build up of organic carbon and nitrogen and C:N ratio (10:1) under vegetation cover. C:N ratio ranged between 5.1 and 6.9 and 3.9 and 10.1, respectively, in surface and sub-surface soil. Nitrogen in soils is present as organic (95-98%) and inorganic but in arid region soils, the inorganic form constitutes about 5 to 18% of the total N. Aggarwal and Lahiri (1981) reported 46 to 67% N as inorganic or mineralized in unstabilized sand dunes of western Rajasthan. Lower organic carbon and available nitrogen have also been reported in the soils of aeolian plains than in other landforms. Total N content of arid soils varied widely, as coarse textured Cirai, Pal, etc. contained 0.028% to 0.050%, medium textured Pipar and Soila (0.042 to 0.056%) and fine textured Asop contained 0.060 to 0.065% N. Aggarwal *et al.*, 1975 reported relatively highest concentration amino acid bound N amongst the various forms in arid soils of Rajasthan.

Phosphorus

Phosphorus present in soils as organic and inorganic forms but organic form constitute hardly 10-20% of the total phosphorus. Total phosphorus content in soils ranges between 300-1500 μg g-1 (Mehta *et al.*, 1971; Deo and Ruhal, 1972; Choudhari *et al.*, 1979) and about 80% of the inorganic P remain bound with Ca. Al-P was higher, generally higher than Fe-P. Total and inorganic phosphorus were irregularly distributed in the soil profile of arid soils, but organic phosphorus decreased with depth (Pareek and Mathur, 1969; Talati *et al.*, 1975). 15-20% (97-110 kg ha-1) of the total P is present in organic form as phytin, lecithin, phospholipids and other unidentified compounds (Tarafdar *et al.*, 1989). The available phosphorus content varies widely in different soils and is about 2.4 to 3.9% of total P and the mean content in different soil series is less than 10 μg g-1 (Mathur *et al.*, 2006). Even though soils are often medium to low in available P, response of P-fertilization in arid soils is generally observed only in good rainfall years (Aggarwal *et al.*, 1989).

Microbial Mobilization of Soil Phosphorus

An extensive survey through arid zones revealed that the population of P-solubilizing micro-organisms is relatively low and varied from one place to place (Venkateswarlu *et al.*, 1984). Bacteria were the dominant group in all the soils followed by fungi. But isolation of actinomycetes especially Streptomyces spp. which can solubilise inorganic phosphates, from arid soil was reported (Rao *et al.*, 1982). However, P-solubilzers isolated from arid soils could not perform satisfactorily in the field because of lack of competition with the native flora. Organic P-compounds must be hydrolyzed by phosphatases of either plant or microbial origin, into plant available form.

Tarafdar *et al.* (1989) reported that activity of phosphatases in different arid soils varied with landuse pattern. High activity of acid and alkaline phosphatases were observed in the rhizosphere of legumes (Tarafdar and Rao, 1990). Tarafdar *et al.* (1988) isolated a number of phosphatase producing fungi (PPF) belonging to Penicillium and Aspergilus from arid soils and their field application enhanced dry matter production and grain yields of clusterbean and mung bean (Tarafdar *et al.*, 1992, 1995).

Arid soils contain 7 to 80 spores of VA-mycorrhizal fungi (Glomus, Gigaspora and Acaulospora genera) 100 g-1 soil. Almost all plants growing in desert including xerophytes do carry VAMF infections on their roots (Kiran Bala *et al.*, 1989) and the infection varied from plant to plant. In general the Young plants mostly carry mycelium with a small number of vesicles while and spores were common on mature roots.

Population of VAM fungi decline during fallow. Crops sown after long periods of fallow have poorly developed mycorrhizal systems and suffer from P-deficiency. VAMF infection on roots of neem was observed up to 250 cm depth, but the intensity of VAMF infection was inversely proportional to the availability of water. VAM-fungi also enhance nodulation and N2-ase activity in legumes (Tarafdar and Rao, 1997). Positive and a significant interaction between *Rhizobium*-VAM fungi was also observed.

Available Potassium

The arid soils are well provided with available potassium (70-890 kg ha-1). The total K content in arid region soils ranged between 980-1890 mg 100 g-1 with an average value of 1489 mg 100 g-1 soil. Major proportion of total potassium in arid soil is present as mineral form followed by interlayer, non-exchangeable and water-soluble form. These forms beside related to each other are correlated with sand and silt fraction and K resistance to depletion (Choudhari and Pareek, 1976).

Arid soils of Rajasthan contain (30 to 1270 mg kg-1) HNO3-soluble and (20 to 1120 mg kg-1) fixed potassium. Some dune and interdunal soils show negative K fixation (Aggarwal *et al.*, 1979; Joshi *et al.*, 1982a). The K fixation capacity was related with the clay content, K-saturation and weathered Kbearing minerals (Talati *et al.*, 1974; Joshi 1986a; Mathur *et al.*, 1981 and Dutta and Joshi, 1993). Kalyansundaram *et al.* (1993) reported only 5.4% soils of north Gujarat as low, 55.2% medium and 38.5% high in available potassium. In the irrigated north-west plain soils of Rajasthan Mathur *et al.* (2006) reported wide variation of potassium content within the soil group i.e. 277 to 488 kg K2O ha-1 in Torrifluvent, 215 to 389 kg K2O ha-1 in Torripsamment and 200 to 374 kg K2O ha-1 in Calciorthid soil group.

Reserve

K constituted 8 to 15% of the total K. K fixing capacity of arid zone soils are lower than of adjoining areas (Choudhari and Jain, 1979). The activity ratio for different arid soils was in the range of 3-21 x 103 mol L-1. Labile potassium, potassium adsorbed on specific and non specific sites, potassium buffering capacities and K potential were generally lower in dune inter-dune and sandy – plain soils than in the medium fine textured alluvial soils (Dutta and Joshi, 1990, 1992; Joshi 1992, 1993). In the soils of Saurastra region water soluble (0.003-0.21 m.e 100 g-1), exchangeable (0.03- 2.0 m.e 100 g-1), non exchangeable (0.32-21.7 m.e 100 g-1) and total (1.1-20.3 m.e 100 g-1) forms of potassium were observed (Patel *et al.*, 1993). Increase in CaCO3 content beyond 30% enhanced the K fixation. Different K parameters, decreased with increase in clay content (Patel *et al.*, 1993).

Secondary Nutrients (Calcium, Magnesium and Sulphur) Calcium is an essential element present in varying amount 0.1 to 5.2% in the arid soil. However, higher amount 15-20% (as CaO) is present in some lower layers of many soils. The exchangeable calcium which is indicator of availability, ranged between 1.5 to 20 m.e. 100 g-1 soil. Fine textured Aridisols of Pipar and Pali contain 15 to 20 m.eq 100 g-1 whereas dune and coarse textured contain between 1.5 to 4.0 and 2.2 to 4.0 m.eq 100 g-1 soil. Magnesium is present in various forms in arid region soil with mineral forms Magnesium is present in various forms in arid region soil with mineral forms exchangeable Fe was 2.2-5.6 ìg g-1 in different arid soils. In the arid soils of Punjab reducible and exchangeable Fe were in ferrous forms, whereas in Rajashthan and Haryana soils it was in ferric form (Shukla and Singh, 1973).

In the arid soils of Rajasthan about 40% soil samples contained 2-5 ppm and 54% samples had 5 to 10 ppm DTPA-Fe (Joshi and Dhir, 1983b). In Haryana soils of Mahendragarh district were well provided, whereas 25% in Sirsa and 51.6% in Hisar district were deficient in iron (Shukla *et al.*, 1975). Iron deficiency reported in Bhatinda (38%), Faridkot (40%) and Sangrur (11%) districts of Punjab. In Jamnagar, Kachchh, Banaskantha and Mehsana districts of Gujarat 8-23% soils were deficient and 45-65% samples marginal in available iron (Dangarwala *et al.*, 1983).

Manganese

Total manganese content in arid soils ranged between 250-875 ìg g-1. Reducible and active forms varied between 3 to 123 on 4 to 128 ìg g-1, respectively (Johri *et al.*, 1978). However, lower contents were reported for Jaisalmer and Barmer soils (Joshi and Dhir, 1982). DTPA extractable Mn in arid soils of Rajasthan varied from 1.1 to 25 ìg g-1 (Dhir *et al.*, 1983). These values were comparable with similar soils of Punjab (Arora and Sekhon, 1981) but the Haryana soils contain higher amount of Mn (6.4-73.8 ppm; Shukla *et al.*, 1975). DTPA extractable Mn in the arid soils of Gujarat showed wide range (0.8-104.4 μg g-1) and only <5% samples were deficient (Dangarwala *et al.*, 1983; Dangarwala and Patel, 1996). In arid Rajashthan 23% samples were below the critical limit, most of which were from the dune and interdune soils (Joshi and Dhir, 1983b). Available manganese was deficient in 29% samples of Hisar and marginal in 40% samples of Sirsa district. The soils of Mahendragarh district were adequate in available manganese.

Zinc

Content of zinc in arid region soils often ranged from 41 to 130 μg g-1. Further, the dune field soils showed higher content compared to Chirai, Khajwana, Gajsinghpur and Pipar soils. Arid soils of Rajasthan are in general well provided with available zinc while deficiency could be encountered only in 13.5 and 9.6 samples in Barmer and Jailsalmer districts, respectively. Wide variations in the contents of exchangeable (0.24-1.28 μg g- 1) and DTPA soluble (0.27-2.36 μg g-1) have been reported (Joshi *et al.*, 1982b; Sharma *et al.*, 1983.

The exchangeable and DTPA soluble forms were not associated with clay, silt and organic carbon. Exchangeable and DTPA soluble forms of zinc showed irregular distribution with depth (Joshi and Dhir, 1981). Sharma and Kolarkar (1983) observed slightly higher values of DTPA Zn (0.83-1.97 μg g-1) in Jamnagar soils.

Dangarwala *et al.* (1983) reported zinc deficiency in 44, 39 and 13% samples in Banaskantha, Jamnagar and Kachchh districts, respectively. The studies revealed that about 80% samples of Hisar, Mahendragarh and Bhiwani districts of Haryana and 25% samples of similar soils in Punjab are deficient in zinc (Anonymous, 1982-83, Arora and Sekhon, 1981). The Zn adsorption by sandy soils follwed the Langmuir adsorption equation.

The quantity of Zn adsorbed by sandy soils was much less than the medium fine textured soils (Joshi *et al.*, 1983b, Joshi and Sharma, 1986). Significant relationships between the quantity, intensity and supply parameters indicated that sandy soils could maintain higher level of available Zn. However, the calcareous soils had low adsorption maxima and high bonding energy constants than the non calcareous soils. Strong affinity of Zn with CaCO3 indicated low availability of Zn in these soils (Joshi, 1996).

Copper

Contents is arid soils varies from 30 to 94 μg g-1 and medium to fine textured Pipar and Pali contained higher amounts than coarse textured Chirai and dune soils. Most of the arid soils of Rajasthan, Punjab and Haryana appeared sufficient in Cu (Dhir *et al.,* 1983; Singh, 1983). However, Takkar *et al.* (1976) reported 22 and 25% samples deficient in Cu in Sangrur and Gurudaspur district respectively. In the Banaskaantha district of Gujarat 22% deficient samples (Dangarwala *et al.*, 1983). The exchangeable (0.23-1.56 μg g-1) DTPA extractable forms of copper showed wide variations in different arid soils (0.28-1.25 μg g-1) (Joshi *et al.*, 1982b; Sharma *et al.*, 1985; Dangarwala *et al.*, 1983).

Micronutrient Transformations and their Relationship with Soil Properties

The transformation of micronutrients in soils depends upon soil pH, soil texture, type of clay mineral, organic matter, salinity/ alkalinity, quality of irrigation water, soil moisture and temperature. In arid soils, the DTPA extractable Fe was significantly and positively correlated with clay, silt and organic carbon and negatively with soil pH (Sharma *et al.*, 1983) and CaCO3 (Joshi *et al.*, 1983b). Most of the Fe deficiency in arid soil is associated with high CaCO3. Total Mn was not related with any of the soil parameters, but HCl soluble Mn was significantly related with clay, silt and organic carbon.

The reducible Mn was not governed by pH and CaCO3 as by clay and organic carbon (Joshi *et al.*, 1981). Free Mn was negatively related with pH and CaCO3 and positively with organic carbon. The mean values of reducible, exchangeable and DTPA soluble Mn decreased with increase in soil pH, CaCO3, organic matter (Joshi and Dhir, 1983a) and finer fraction of soils (Joshi and Dhir, 1983b, Lal and Biswas, 1973). However, for Zn, only HCl soluble fraction was associated with clay, silt and organic carbon, but exchangeable and DTPA soluble forms were not related with any of the soil parameters (Joshi *et al.*, 1982b; Sharma *et al.*, 1983). The HCl soluble Cu was significantly related with clay, silt, organic carbon and CaCO3 content of soils. Available Cu was significantly related positively with clay, silt and organic carbon and negatively with pH. Exchangeable Cu was critical in arid brown and marginal in sierozem soils of Punjab and Haryana, respectively (Grewal *et al.*, 1969; Singh and Shukla, 1984; Singh *et al.*, 1988). Medium textured soils contained slightly higher amounts of DTPA soluble Cu than dune, interdune and sandy plain soils.

There was marked effect of saline/high RSC irrigation water on availability of micronutrients in soils. Soils irrigated with saline and high RSC waters were low in available micronutrients (Joshi, 1988, 1990b). The contents of HCl soluble, free and reducible forms of Mn in salt affected soils were comparable with the associated non saline soils, but their predictability was much reduced (Joshi *et al.*, 1988a). However, little difference was observed in HCl-Fe and free Fe in salt affected and associated non saline soils. Due to the effects of different soil parameters in salt affected soils, the HCl-Fe was better predictable but free Fe had low predictability (Joshi *et al.*, 1988a). Exchangeable Fe in salt affected and associated non saline soils were comparable. The salt affected and associated non saline soils did not differ in various forms of Zn, but high salinity appears to vitiate their predictability (Joshi *et al.*, 1988b). Though there was no difference in the contents of HCl soluble and exchangeable Cu in salt affected soils and the associated nonsaline soils, but their predictability was considerably low (Joshi *et al.*,1988b). Adsorption/desorption behaviour of micronutrients in arid soils.

The adsorption of Zn and Cu in arid soils followed Langmuir adsorption equation. Because of the low clay/organic matter, low cation exchange capacity, high base saturation and calcareous nature of sandy soils, they have characteristic adsorption and release behaviour for these nutrients. The quantity of Zn adsorbed by sandy soils was much less than medium/fine textured soils (Joshi *et al.*, 1983b; Joshi and Sharma, 1986).

Significant relationship between quantity, intensity and supply parameters indicated that sandy soils could maintain higher level of available Zn. Strong affinity of Zn with CaCO3 indicated low availability of this nutrient in such soils (Joshi, 1996). However, Cu could be adsorbed at low concentration but with increasing addition, there was precipitation of Cu rather than multilayer adsorption (Joshi, 1995). The adsorption of Cu was less in fine sandy than loamy soils (Joshi, 1986b).

Micronutrients in Plants and their Recycling in Arid Soils

It is possible to estimate the extent to which the micronutrient cations are recycled through the decomposition of plant materials from assumptions about the dead materials and the concentration of cations. Rainfed crops and grasses/shrubs of arid region have higher amount of micronutrient content as compared to irrigated crops. Among grasses, substantial higher amount of copper was observed in bharut (*Cenchrus biflorus*) and sewan (*Lasiurus sindicus*) grass. Zn was found deficient in irrigated crops (groundnut, wheat, and mustard), but was not the case with rainfed crops (Soni *et al.*, 2003).

Among the perennial components, the return of Fe is maximum through trees (Total Fe=381 ppm) followed by shrubs (Total Fe=251 ppm) and grasses (Total Fe=235 ppm) of arid region (Sharma *et al.*, 1985). As regards Mn, arid zone grasses have highest concentration of Mn followed by tree species. Grasses have also significantly higher values of Zn and Cu as compared to trees and grasses (Sharma *et al.*, 1984). Fleming (1973) and Dhir *et al.* (1985) also stated that arid zone grasses are significantly low in Mn and quite high in Zn and Cu.

Various soil degradation processes viz., Aeolian hazards, water erosion including floods have detrimental effect on micronutrient availability (Raina and Joshi, 1994).The soils under natural pasture were low in available micronutrients than the soils under sown ungrazed pastures (Raina and Joshi, 1991). Comparatively higher amount of micronutrient was observed in natural rangelands followed by arable rain fed cropping and minimum in irrigated conditions (Soni *et al.*, 2006).

Considering the status of natural rangelands as ideal, the percent losses of Fe, Zn, Cu and Mn showed that reduction in available micronutrients was more in irrigated lands as compared to rain fed cultivated lands. On an average, Fe, Zn, Mn and Cu were reduced by 32.4, 24.0, 23.8 and 12.0% in irrigated and 9.2, 8.4, 10.4 and 6.8%

in rain fed cultivated soils over natural rangeland systems, respectively (Soni *et al.*, 2006). Build up of all the four micronutrients (Mn, Zn, Cu and Fe) was more in medium and fine textured soils under *Oran* system as compared to cultivated lands.

Nutrient Management in Arid Soils

Nutrient use efficiency is evaluated using metrics that reflect crop uptake of fertilizer added in the current growing season. The problem of low nutrient use efficiency is conventionally viewed as a consequence of temporal asynchrony and spatial separation between applied nutrients and the crop. As a result, efforts to improve nutrient utilization efficiencies have emphasized on improved delivery of nutrients to the root zone during the period of crop uptake through modifications such as banding, fertigation, split fertilizer applications, etc. To increase crop access to N fertilizer, a variety of additives have been developed that inhibit urease activity, nitrification and denitrification. These approaches have been extremely successful in terms of increasing nutrient use efficiency and maximizing yields; however, their utilization on mass scale has met with limited success.

Despite more than 30 years of concentrated effort, mass balances indicate that annual N and P inputs consistently exceed harvested exports by 40 to >100% resulting in substantial losses of these nutrients to the environment. Thus, it is evident that global challenge of meeting increased food demand and protecting environmental quality will be won or lost in cropping systems (Cassman *et al.*, 2002). Major limitation of most of the approaches tried in past three decades is lack of their applicability on large scale on small farm holdings. Only two approaches hold promise for application on mass scale to increase nutrient use efficiency (1) application of optimum quantity and (2) internal nutrients circulation.

Application of Optimum Quantity

One of the first step necessary for achieving higher fertilizer use efficiency (FUE) is to apply the right amount of fertilizer. But the long term estimations of N requirement of crops under rain-fed conditions may be very difficult as the yield levels vary from zero to three times of average yield. Using average yields in arid regions is too conservative and may actually result in lowering average yields with time because of insufficient nutrient availability for the very favourable years and on contrary, use of relatively high yield goals results in excess N applications in most years and can greatly reduce profit.

The current concern over the potential for excess N to degrade the environment also makes this alternative unacceptable. Tucker (1988) presented three ground rules to arrive at logical yield goals. These include choosing yield goals based upon: (1) highest yield within the past 5 years with good crop management (2) yield goal set a 1.5 times of long term average, and (3) yield goal based on soil capabilities as defined in Standard Soil Surveys, using yields of highest yield obtaining farmer's in the vicinity on the same kind of soil.

Uncertainties in soil nitrogen (N) supply in rainfed agriculture and crop N demand present a challenge to scientists in deciding on N fertilizer rates. Number of field studies have documented that the improvements in N use efficiency is possible with site-specific N management approaches. Lobell (2007) has presented a general model of N rate decision-making which computes the optimal N rate that maximizes expected profit given uncertainties in N supply and demand.

The cost of uncertainty is measured as the difference in N rate when soil N supply and crop N demand are unknown versus known perfectly. Eliminating uncertainty in soil N supply (but not crop demand) reduces average N rates by 5–40% in different crops while perfect knowledge of potential crop N demand (but not soil supply) reduces rates by 3-10%. Simultaneous knowledge of both factors reduced N rates by significantly more than the sum of their individual effects. This approach is very different from most extension services in India which provide a single, standard fertilizer recommendation for an entire district or region. Farmers apparently have few guidelines for adjusting N-fertilizer amount to account for the large differences in the indigenous N supply, indicating the need for a 'field-specific' approach to N management. Fieldspecific N management could lead to substantial reductions of N rates without yield loss in a wide range of cropping systems, thereby improving profitability and environmental quality.

Internal Nutrient Circulation

The approach comprises of various conventional practices like recycling of crop residues, either through direct application or as manure and cultivation of legumes often referred as integrated nutrient management in past. The practice of effective use of inorganic and organic sources of nutrients together in a proper proportion not only reduces the requirement of inorganic fertilizers, but also improves physical conditions of soil, enhances water retention capacity and its availability in the soil. Apart from this, the biological properties of soil are also improved considerably (Sharma *et al.*, 2005, 2008, 2009) and fertilizer use efficiency is also increased.

The results of several studies have shown that the combined use of fertilizer and farm yard manure (FYM) is very much essential in rainfed areas for sustaining moderate to high yields and achieving higher nutrient use efficiencies. Besides, application of FYM and composted organic wastes and other organics improve yield stability in rainfed areas (Aggarwal and Venkateswarlu, 1989; Venkateswarlu and Hegde, 1992; Singh *et al.*, 2000). Singh *et al.* (1981) observed that under arid conditions of Jodhpur continuous application of sheep manure in general gave substantially higher yields than the application urea alone. Rao and Singh (1993) showed that substitution of 50% of fertilizer requirement by FYM resulted in yield levels nearly similar to those obtained with complete fertilization.

Aggarwal and Praveen-Kumar (1996) on the basis of a seven year study on arid soils showed not only a beneficial effect of FYM application but also a synergistic effect of simultaneous application of FYM and inorganic fertilizers on crop yield. FYM application increases the utilization efficiency of fertilizer N, however, improvement in soil fertility after FYM application is a very slow process. Nutrient content of pearl millet, sorghum stover is relatively low, stover can contribute to the productivity of the soil. Such residues must be managed carefully, however, because N can be immobilized at the time of peak maize N requirements, resulting in poor crop growth. In rainfed areas, cereal stover is often fed to livestock, and manure is applied in field. This way of recycling the residues is more beneficial for crop than their direct application in field. Losses of N from such systems are often high. Cattle manure is applied in a dried, aerobically decomposed form, often with a high sand content and N content that is frequently around 1% or less.

Research shows that the most efficient use of manure is to combine it with some inorganic fertilizer. Ladd and Amato (1985) and Snapp and Silim (2002) showed that the most promising route to improving inorganic fertilizer efficiency in cropping systems is by adding small amounts of high-quality organic matter (possessing a narrow C/N ratio and a low percentage of lignin) to soils. It provides readily available N, energy (carbon), and nutrients to the soil ecosystem, and improve structure, and increase soil microbial activity and nutrient cycling and reduce nutrient losses from leaching and denitrification (De-Ruiter *et al.*, 1998). Soil microbes are valuable not only because they supply nutrients directly, but because they enhance the synchrony of plant nutrient demand with soil supply by reducing large pools of free nutrients (and consequent nutrient losses from the system). Thus, microbes maintain a buffered, actively cycled nutrient supply.

Incorporation of crop residues and natural vegetation in soil improve microbial activity during decomposition. Also adhesive action of decomposed products improves soil aggregation, hydraulic conductivity and moisture retention (Venkateswarlu, 1984, 1987, Gupta, 1984; Gupta and Gupta, 1986). Leaving the crop residues in soil generally have a positive effect on grain yield (Aggarwal *et al.*, 1996). Crop residues are also efficient source of nutrient like other organics viz., cattle manure and compost. However, Aggarwal *et al.* (1996) did not find any significant change in the yield of succeeding crop of pearl millet after addition of crop residues with wide C:N ratio, whereas it was significant after incorporation of residues with narrow C:N ratio.

Legume rotations are an important practice for restoring soil fertility on larger land holdings. The amount of N returned from legume rotations depends on whether the legume is harvested for seed, used for forage, or incorporated as a green manure. Estimated net N return of 23-110 kg ha-1 from pigeon pea, 23-50 kg ha-1 from dolichos beans, and 25-60 kg ha-1 from groundnuts has been reported by Giri and De (1980). Mishra (1971), Mann and Singh (1977), Singh (1980) and Singh *et al.* (1985) also reported that in arid soils pearl millet-cluster bean rotation gave higher yield than of continuous cultivation of pearl millet due to improved soil fertility (Praveen-Kumar *et al.*, 1996). Singh *et al.* (1985) in a long-term study found an increase soil organic carbon by 12% and available soil P by 25% after legume cultivation. Singh and Singh (1977) on the basis of a longterm study reported that cultivation of green-gram in rotation with pearl millet supplied with 20 kg N ha-1 gave similar yield as with application of 40 kg fertilizer N ha-1. Singh *et al.* (1985) observed that rotation of pearl millet with green gram or clusterbean was better than its rotation with moth bean.

Praveen-Kumar *et al.* (1996) reported that higher yield of pearl millet when it was preceded by clusterbean cultivation than green gram. Beneficial effect of legumes to pearl millet also depends on the number of seasons of their cultivation prior to pearl millet. The intercropping of pearl millet and legumes in arid soils has also shown promising results (Mishra, 1971; Singh and Joshi, 1980; Singh *et al.*, 1978).

Major quantity of N i.e, upto 70% of total N uptake, can be taken up by the pearl millet within first 30 days of growth. Though it slows down thereafter, but continues gradually till the grain filling stage. This high N demand cannot be met from the soil and thus pearl millet respond favorably to the N addition (Aggarwal and Vekateswarlu, 1989), which ranged from 7.0 to 18.0 kg grain kg-1 N with higher values in good rainfall years. Singh *et al.* (1981) have reported that the yield of pearl millet doubled with the application of 40 kg N ha-1. Similar results have also been reported by Singh *et al.* (1981). Aggarwal and Praveen-Kumar (1996) on the basis of a seven year long study reported significant response to application 80 kg N ha-1 only in the years of good rainfall. Comparison of different N fertilizers showed that, maximum yields were recorded with ammonium sulphate. But due to high cost of ammonium sulphate, urea has become the major source of N in arid regions even though utilization efficiency of N from urea by pearl millet is very low. Various studies in CAZRI Jodhpur have revealed that the mixing of elemental S with urea (Aggarwal *et al.*, 1987) or application of small quantity of ammonium sulphate before the application of urea (Praveen- Kumar and Aggarwal, 1988) increases its efficiency. Praveen-Kumar and Aggarwal (unpublished data) also showed that NH4 : NO3 ratio in a N fertilizer also effect its utilization efficiency and observed maximum yield levels and highest utilization efficiency from the applied N fertilizer when NH4 : NO3 ratio in the fertilizers was 3:1. Besides ammonia volatilization denitrification is now being considered an important pathway of N loss (Peterjohn, 1991). Due to different loss mechanisms major part of N applied, if not used by crop is lost from soil. As a result residual effect of fertilizer N is generally not observed (Aggawal and Praveen-Kumar, 1996). Praveen- Kumar and Aggarwal (Unpublished data) demonstrated this in a field experiment wherein 20, 40, 60 and 80 kg N ha-1 was applied in different plots at the time of sowing and the crop was harvested after 20, 40 and 60 days of growth symbolizing complete crop mortality at early, medium and late stage of drought, respectively. Next year pearl millet was again grown in these plots without addition of N fertilizer to study the residual effect. The residual effect was compared with the yield in some other plots where fertilizer @ 0, 20 and 40 kg N ha-1 was added. Results indicated comparable grain yield in most plots where residual effect was studied to those in control plots where no fertilizer was added. This suggested that fertilizer N applied previous year did not show significant residual at any level of fertilizer application and stage of crop harvest. Since droughts occur frequently in arid regions because of which crop often fails, therefore, farmers consider application of N fertilizers a risky input and hence the need for its better management. Split application of N fertilizers is therefore, considered a risk avoiding strategy which results in the higher yield as compared to their one time application, increased utilization efficiency and lower risk. Praveen- Kumar and Aggarwal (1997, 1998) observed basal application of fertilizer N may be avoided if clusterbean has preceded the pearl millet.

Biodiversity in Soils

Soil biodiversity is an important resource that provides ecosystem processes essential to the functioning of natural and global systems. Biota occupy remarkably different habitats within profile of arid soils with some species living on decaying biological matter, some in the rich organic matter horizon, and some in deeper layers and the species composition may vary even at millimeter scale depending on type and amount of vegetation and soil texture. Soil organisms moderate numerous ecosystem processes (e.g., decomposition, C and N transformation, hydrologic cycles). Changes in landuse, climate and atmospheric can alter species composition within the soil community, and may impact ecosystem function. Current priorities in the study of soil biodiversity include:

(1) understanding the functions of rarely sampled fauna so that species estimates and global distributions are based on an accurate and current database;

(2) determining which habitats are most vulnerable for soil biodiversity loss, e.g., where the 'hot spots' of biodiversity are, and which habitats and what time frames are most amenable to restoration;

(3) synthesizing data and determining which invertebrate and microbial 'species' are key to ecosystem processes;

(4) identifying gaps in knowledge of multispecies interactions and their influence on ecosystem functioning;

(5) collaborating on long-term (more than 3 years) experiments to examine effects of global changes on aboveground-below-ground biodiversity linkages and ecosystem functioning; and

(6) determining, based on natural history information, which species are more likely to be invasive if introduced, and using this potentially to reduce spread and to identify threats to other species.

Macro-fauna

The arid soils are home to diverse life forms adapted to hot and dry milieu. Soil organisms contribute in formation of organic matter and nutrient cycling. The burrowing invertebrates create macropores that allow rapid flow of water into or through soil. The tiny arthropods produce fecal pellets that are mixtures of soil and organic matter. These become stable soil aggregates. The arid soil meso and macro fauna are constituted largely by insects, mites, earthworms and millipedes. Predators in the soil food web include scorpions, centipedes, spiders, mites, some ants and beetles. They check the population of many soil biota. The reptiles and rodents represent the soil dwelling vertebrates, the former play a predatory role while the latter thrive on a variety of food resources.

Soil compaction, low vegetation or plant litter reduces the number of soil arthropods. Their level of activity depends on micro-environmental conditions including temperature, moisture, aeration, pH, pore size, and types of food sources. Earthworms move soil from lower strata up to the surface and move organic matter from the soil surface to lower layers. Suthar (2002) studied the earthworm resources of arid Rajasthan and their utilization in vermitechnology. Suthar (2009) recorded occurrence of nine species of earthworms from Jodhpur district. These included *Pontoscolex corethrurus,* Amynthas morrisi, Metaphire posthuma, Lampito mauritii, Perionyx sansibaricus, Ocnerodrilus occidentalis, Dichogaster bolaui, Ramiella *bishambari* and *Octochaetana paliensis*. Diversity and abundances of earthworms in conventional, integrated and organic agro-ecosystems was also measured. Maximum numbers of earthworm were found in integrated farming system and abundance of earthworms in arable system was directly related to the management practices (Suthar, 2009). There have been but few studies on the soil macro faunal associations in different production systems of the arid regions and their contribution to soil fertility. Tripathi *et al.* (2005) reported that Acari, Myriapoda, Coleoptera, Isoptera and Collembola arthropods as the major soil faunal groups in silvipastoral systems of arid zone. Relative densities of Acari, Myriapoda and other arthropods were highest in silvopastoral systems and those of Coleoptera, Isoptera and Collembola were highest in pure grass plots. Termites are another important arthropods in the soil.

Termites aid in crop residue breakdown, enhance the porosity of soil, increase the soil's infiltration capacity and affect nutrient cycling by concentrating organic matter in the nests. However, their beneficial role is shadowed by the damage they cause to the cultivated plants. All the known species in the arid regions are subterranean. This group of arthropods has been studied fairly well owing to their economic importance in all fields. Nearly two scores of termite species have been recorded from the arid region (Rathore, 1996). Parihar (1981) collated the available information about termites and their management.

The known termite fauna of the arid parts of Rajasthan, Gujarat, Haryana and Punjab are represented by the following species: *Amitermes belli, Anacanthotermes macrocephalus, Angulitermes* jodhpurensis, Coptotermes heimi, C. kishori, Eremotermes fletcheri, E. neoparadoxalis, E. paradoxalis, Heterotermes indicola, Incisitermes didwanaensis, Microcerotermes cameroni, M. heimi, M. laxmi, M. palestinensis, M. raja, M. sakesarensis, M. tenuignathus, Microtermes mycophagus, M. obesi, M. unicolor, Odontotermes assmuthi, O. bellahunisensis, O. brunneus, O. dehraduni, O. distans, O. feae, O. giriensis, O. girnarensis, O. guptai, O. gurdaspurensis, O. latiguloides, O. microdentatus, O. obesus, O. paralatiguloides, O. redemanni, O. sasangirensis, O. O. wallonensis, O. lokanandi, O. indicus, Psammotermes rajasthanicus, Speculitermes cyclops, S. sinhalensis, Synhamitermes *quadriceps, Trinervitermes biformis* and *T. fletcheri.*

Ants also modify soil properties through their nesting and foraging activities. While most of the ants found in arid regions are foragers, a few are predatory and some tend sucking insects. Still there have not been extensive studies about the foraging by ants in the arid regions. The available information is mostly restricted to documentation of the occurrence of ants in the arid tracts (Tak and Rathore, 1996; Tak, 2008), which include *Acantholepis frauenfeldi, Anochetus punctiventris, A. sedilloti, Camponotus* angusticollis, C. compressus, C. irritans, C. mitis, C. taylori, Cataglyphis *bicolor setipes, Crematogaster brunnea contemta, Dorylus labiatus, D. orientalis, Meranoplus bicolor, Messor barbarus humalayana, Monomorium* atomus, M. criniceps, M. destructor, M. glaber, M. indicum, M. pharaonis, M. scabriceps, M. wroughtoni, M. gracillimum, M. latinode, Pheidole sulcaticeps, *P. wroughtoni, Sima rufonigra* and *Tapinoma melanocephalum.*

Coleoptera is another important order of insects associated with soil. The grubs of most beetles are inhabitants of soil, plant roots or decaying matter. Some of these are also predators of injurious insects. Being of agricultural importance, beetles and weevils have received attention of researchers. However, there is no account of their benign contribution in enriching the soil. Pal (1977) had given an account of the white grubs prevalent in the arid regions. Tripathi *et al.* (2008) studied the diversity and abundance of beetles in some silvipastoral systems. The more common Scarabaeids in the arid region are species of the genera *Lachnosterna* (*Holotrichia*), *Anomala*, *Adoretus*, *Rhinyptia*, *Schizonycha*, and *Onthophagus*.

Bembidion luniferum, *Asaphidion triste*, *Leitus indus* and *Scaritus* species are more common Carabids. There is a large number of other beetles whose grubs inhabit the arid soil. The grubs and adults of dung beetles contribute profusely in decomposing the organic matter in arid soils. The beetles process large amounts of animal dung into nodules or balls, and roll these into subterraneous chambers or tunnels where they are degraded, thereby increasing soil fertility. They also prevent the use of decaying materials by other pest insects. Eighty-five species of these beetles have been documented from the arid regions belonging to *Onthophagus*, *Caccobius*, *Onitis*, *Copris*, Scarabaeus, Catharsius, Gymnopleurus, Oniticellus, Phallops, Drepanocerus, *Heliocopris* and *Chironitis* genera (Ram-Sewak, 2009).

The Collembolan insects contribute significantly in litter decomposition. The majority of springtails feed on fungal hyphae or decaying plant material. But not enough information is available about them from the Indian arid soils. Seven species of Collembola including *Xenylla obscura* and *Isotomodes dagamae* were recorded from Sriganganagar district by Faisal and Ahmad (2005). Their existence and establishment has been attributed to the change in land use pattern due to the development of canal irrigation. Although a large number of species of soil mites exist in the arid regions, there have not been any systematic attempts to study these organisms.

Sanyal (1996) while giving an account of the soil mite fauna from Rajasthan has listed only three species *viz.*; *Eugamasus* sp, *Galumna flebellifera* and *Microsmaris pinnata* collected from the drier tracts. There have been many studies on rodents in the arid regions, their damage and control strategies, but no precise estimates are available about their contribution to soil fertility, if at all. The soil inhabiting rodents in the arid zones include *Bandicoota bengalensis*, *Gerbillus gleadowi*, *G. nanus*, Meriones hurrianae, Mus booduga, M. cervicolor, M. musculus, M. platythrix, *Nesokia indica*, *Rattus gleadowi*, *R. meltada pallidior* and *Tatera indica*. There is a need to generate more information about the soil meso and macro fauna that influence soil fertility under arid conditions and the ways to enhance their activities.

Soil Microorganisms

Microbial activity in desert soil is highly dependent on characteristics such as temperature, moisture and the availability of organic carbon (Parkar *et al.*, 1984; Rao and Tarafdar, 1998). Of these, moisture availability is the major factor affecting microbial diversity, community structure and activity. Populations of aerobic bacteria in deserts across the world are reported to vary from <10 in Atacama desert to 1.6´107 g-1 in soils of Nevada (Skujnis, 1984). Populations in Indian arid zone reported to have relatively smaller (1.5´102-5´104 g-1 soil) (Venkateswarulu and Rao, 1981; Kathuria, 1998). Gram-positive spore formers are dominant and the populations do not decline significantly even during summers (Rao and Venkateswarlu, 1983).

Actinomycetes may constitute ~ 50% of the total microbial bacterial population in desert soils. Dominant microflora of desert soils is made up of coryneforms, i.e. *Archangium*, *Cystobacter*, *Myxococcus*, *Polyangium*, *Sorangium* and *Stigmatella*; sub-dominant forms comprise *Acinetobacter*, *Bacillus*, *Micrococcus*, *Proteus* and *Pseudomonas*. Cynobacteria also contribute significantly (0.02 to 2.63 ´ 104 g-1 soil) to the biota of hot arid regions in terms of primary productivity and nitrogen fixation (Bhatnagar *et al.*, 2003). The dominant cynobacterial forms of Thar desert are *Chroococcus minutus*, *Oscillatoria pseudogeminata* and *Phormidium tenue*; *Nostoc* sp. dominates amongst heterocystous forms. Fungal populations as viable propagules range from 0.5 to 14.7´103. The dominant genera include *Aspergillus*, *Curvularia*, *Fusarium*, *Mucor*, *Penicillium*, *Paecilomyces*, *Phoma* and *Stemphylium*. Xeric mushrooms such as *Coprinus*, *Fomes*, *Terfezia* and *Teramania* and arbuscular mycorrhizal fungi such as genera *Glomus*, *Gigaspora* and *Sclerocystis* have also been reported from desert (Trappe 1981; Pande and Tarafdar, 2004). Rhizosphere is an important site of microbial activity in desert soils, since it provides ample carbon substrate in an otherwise organic matter poor arid soil. Generally R:S (Rhizosphere : soil) ratio is high in arid soils for nearly all metabolic types of bacteria (viz. heterotrophs, dizotrophs, cellulites and nitrifiers) and fungi in most plants studied (Kathuria, 1998; Singh and Tarafdar, 2002). The soil zone penetrated by fine roots and held together by mucilage is called the rhizosheath.

Bacillus polymyxa and the fungus *Olpidium* are found associated with rhizosheaths. *Ancalomicrobium* and *Hypho-microbium*- like organisms are also present (Wullstein and Pratt, 1981). Rhizosheaths are important because of the associated diazotrophs and enhanced water retention and nutrient uptake (Watt *et al.*, 1994). In the rhizosphere of desert plants mycorrhiza play a very significant role in plant nutrition and stabilization (Tarafdar and Rao, 1997, 2002; Tarafdar and Gharu, 2006). Most species in families of *Asteraceae, Fabaceae, Poaceae, Rosaceae* and *Solanaceae* usually form endomycorrhizal associations in arid habitats (Skujins, 1984). Trappe (1981) listed 264 plant species from arid and semi-arid environments that had mycorrhizal-based root colonization and approximately 25% species exhibited specific associations with endomycorrhizal fungi. *Glomus deserticola* is indigenous to many desert soils (Bhatnagar *et al.*, 2003). Kiran Bala *et al.* (1989) reported >50% infection by vesicular-arbuscular mycorrhizal (VAM) fungi in 17 tree species of the Indian desert, with genera *Glomus* and *Gigaspora* being dominant.

Opunitia and *Euphorbia* showed considerable root infection. In desert, incidence of arbuscular mycorrhiza (AM) infection varies with the availability of water (Tarafdar, 1995) and with composition of plant community (Kiran Bala *et al.*, 1989). Mycorrhiza also help in desert reclamation and soil stabilization. They link soil particles to each other and to the roots in part by producing glomalin, an important glue that holds aggregates together. Panwar and Vyas (2002) reported beneficial effects of *Acaulospora mellea, Gigaspora margarita, G. gigantean, Glomus deserticola, G. fasciculatum, Sclerocystis rubiformis, Scutellospora calospora* and *S. nigra* on *Moringa concanensis* and proposed use of such AM fungi in conservation of this endangered multipurpose tree species in Indian desert. Crusts on soils are formed by microorganisms and microphytes. These are variously called as cryptogannic/cryptobiotic/biological/cynobacterial/ microphytic crusts (West, 1990). A consortium of green algae, cyanobacteria, lichen, fungi, bacteria, diatoms, mosses and liverworts form cryptobiotic crusts. Mosses and liverworts seem to favour the comparatively more mesic sites. Cynobacteria favours the harshest sites, and lichens dominate in intermediate sites.

Cynobacterial crusts generally dominate poor sandy soils. Lichens increase proportionately with carbonates, gypsum and silt content of the substrate (Gracia-Pichel *et al.*, 2001; Budel, 2002). *Microcoleus chthonoplastes* is dominant in saline sand crusts (Gracia-Pichel *et al.*, 2001). The most common genera are *Microcleus* (*M. Chthonoplastes, M. paludosus, M. sociatus* and *M. vaginatus*) and *Nostoc* sp. Other forms are *Calothrix, Lyngbya, Oscillatoria, Phormidium, Scytonema* and *Tolypothrix*. Common green algae of soil crusts are *Chlorella, Chlorococcum, Coccomyxa* and *Klebsormidium*. Dominant fungal genera in crusts are *Alternaria, Fusarium* and *Phialomyces* whereas in non-crusted soils, *Alternaria* and *Penicillium* dominate followed by *Fusarium*. Approximately 90% of the aerobic bacterial population of cryptobiotic crusts consists of corneform bacteria (Skujins, 1984).

Diversity status of cryptobiotic crusts at Thar desert in India showed 43 morphotypes of diazotrophs in BG11-N enrichment and 71 of algae and cyanobacteria in the same medium supplemented with nitrogen (Bhatnagar *et al.*, 2003). Most frequent form was *Phormidium tenue*. In the case of diazotrophs the most frequent forms were *Nostoc punctiforme, Nostoc commune* and *Nostoc polludosum*. Cynobacteria presence was influenced by their plant partners. *Alternaria* sp. was the dominant fungus in these crusts. Microorganisms in soil are poorly characterized. It has been established that the genetic diversity of soil has an estimated 6400 to 38000 prokaryotic species per gram. On the other hand, less than 0.3% of the microorganisms present are culturable by standard techniques (Pettit, 2004). Only a handful of studies have looked at microorganisms in tropical soils. Microbiological properties of these soils also varied depending on the type of land use patterns. Grasslands, in general, supported higher number of microorganisms than tree plantations, cultivated fields or barren land. Stabilization of shifting and sand dunes introduction of vegetation has markedly increased the soil microflora. In general, the low organic matter content and poor moisture availability of desert soils were the major factors limiting optimum microbial activity.

Fungi like *Fusarium, Gliocladiu, Penicillium* and *Trichoderma* are stress tolerant. Majority of fungi are mesophiles with maximum growth between 25 and 30°C (*Mucor mucedo, Mortierella, Penicillium chrysogenium*), however, *Cylindrocarpon* sp. *Candida scotti* are cold tolerant (psychrotolerant) and can grow near 0°C; others are thermotolerant and grow above 40°C (*Rhizomucor, Thermomyces, Talaromyces*). Xerotolerant fungi can grow on dry material (*Aspergillus, Pencillium*) with low atric potential (aw) while osmotolerants grow at very low osmotic potential (*Pichia* sp.). Dung of herbivorous mammals harbors a large number of fungi, termed coprophiles, of while *Pilobolous, Ascobolus* and *Basidiobolus* are famous for their special shot-gun dispersal mechanism ! An interesting ecological group of fungi captures and grows parasitically on nematodes, their cysts and eggs (nematophagous); over 150 species are reported (Gray, 1998) within the genera *Catenaria, Dactyela, Harposporium, Monacrosporium, Nematophthora, Rhopalomyces* and

Stylopage. Such fungi are good trapping agents in the biological control of nematodes.

Potentials of the Arid Zone Soils

Soils in the drier regions have low reserves of soil organic matter. Singh *et al.* (2007) reported 2.13 Pg carbon stock in the 0-100 cm depth of which 1.23 Pg as soil organic carbon and 0.90 Pg as soil inorganic carbon, in the arid and semi arid regions of Rajasthan. They had studied the same under different soil types. Per year 4200- 4600 kg km-2 soil organic carbon can be sequestered in the untilled soils of arid Rajasthan, if they are put under canopy cover round the year. The desert and saline soils in the Kutch peninsula is reported to have a current total carbon stock of 0.8 Pg with an estimated potentials of 1.3 Pg carbon. Uncertainties persist in the estimation of net flux of CO_2 from the soils of tropical areas due to inconsistency in land use/ land cover pattern (Chapin and Ruess, 2001; Jones *et al.*, 1998). The studies conducted at RRS, Kukma indicated soil organic carbon stock of 35.13 t ha-1 under the silvipastoral system involving *Neem* and *Cenchrus ciliaris,* 31.9 t ha-1 under Acacia and *Cenchrus ciliaris,* in the 100 cm soil depth. The carbon sequestration in soil is strongly affected by the root production (Matamala *et al.*, 2003). The trees in the arid zones put forth a large volume of below ground biomass in the form of roots. The tree roots play an important role in adding organic matter to the soil. About 25% of the total living biomass of trees is in roots which continuously add organic matter to soil by death and decay of roots. It is more important in arid zone where climax tree species like *Prosopis cineraria* that have very deep root system which can reach upto 70 m depth. Thus in arid zones large volume of carbon gets sequestered in lower layers, that have high resilience. Narain (2008) reported planting trees and grasses in the degraded lands of arid zone can help to increase the soil carbon stock from 24.3 Pg to 34.9 Pg. The studies conducted at RRS, Kukma indicated a total below ground carbon stock of 1.61 t ha-1 that constituted 23.4% of total carbon stock under silvipastural system involving *Acacia tortilis* and *Cenchrus setegerus* and 27.6% contribution by below ground portions to the total carbon stock under *Acacia tortilis* and *Cenchrus ciliaris*. The intentional controlled burning of biomass can produce charcoal or black carbon that can contribute to carbon sequestration (Goldberg, 1985). The activated charcoal, apart from locking carbon in the soil for centuries, can attribute positively to the cation exchange capacity, physical, chemical and biological properties of soil.

10

Prosopis Species in the Arid and Semi-Arid Zones of India

The genus *Prosopis* contains 44 species in five sections, spread throughout southern Asia, Africa and America, but predominantly in the latter continent, from western North America to Patagonia. The centre of polymorphism is in central-western Argentina. *Prosopis* species are generally thorny and frequently phreatophytic. Plants exhibit an enormous variety in growth habits and in their defences, which may be represented by thorns, internodal prickles or be entirely absent.

In the Indian sub-continent, only one species, *P. cineraria* is endemic, all others having been introduced. *P. juliflora* which now occurs throughout the arid and semi-arid tropics of the Indian sub-continent, was first introduced last century. In India, the species now grows predominantly in Rajasthan, Haryana, the plains of Uttar Pradesh, Gujarat, North Karnataka, Tamilnadu, Madhya Pradesh, Delhi, Maharastra and also in parts of Punjab. Depending on climatic and edaphic conditions, the density of species varies from 5 to 100 trees/ha.

The growing of arable crops in association with *P. cineraria* is an age-old tradition in arid parts of India in general, and particularly in the Thar desert. The vegetation of the Thar desert region is classified as dry tropical thorn forest (Champion and Seth, 1968) and in a broader successional gradient, *P. cineraria* is designated as the climatic climax species, and is the predominant constituent of the vegetation complex of arid parts of India. In its natural range, the species grows very slowly, especially during early growth.

Realising the slow growing nature of *P. cineraria*, another *Prosopis* species, *P. juliflora* was first introduced into India in 1857 from Latin America. The species reflected its potential in terms of adaptability and growth in these new environmental conditions immediately after introduction. The arid and semi-arid parts of the country were the most preferred habitats of *P. juliflora*, but over time this species has spread into other parts of the Indian sub-continent. *P. juliflora* is now well naturalised in the arid and semi-arid tracts of the region.

Introduction and distribution of *Prosopis* in India

Although the earliest recorded introduction of *P. juliflora* was in 1857, the first systematic plantations were not carried out until 1876 in the Kaddapa area of Andhra Pradesh. It was introduced into parts of Gujarat in 1882 and into Sindh (Pakistan) in 1877. The ruler of the princely state of Marwar introduced *P. juliflora* into Rajasthan in 1913. Fascinated by its excellent growth rates in the region, the then ruler of the state declared it a "Royal Plant" in 1940 and exhorted the public to protect it. Since then, the species was gradually introduced in many parts of the country including Haryana, Punjab, Uttar Pradesh, Madhya Pradesh, Maharastra and Tamilnadu. Due to its wide ecological amplitude and rapid colonising ability, the species has spread rapidly over a large part of the arid and semi-arid tract, with the exception of frost prone areas.

At present, *P. juliflora* is thought to inhabit more than 500,000 ha, although plant densities vary with different areas and habitats. It grows in all type of soils, from sandy to saline-alkaline soils. It has proved to be a most versatile species for afforesting shifting sand dunes, coastal dunes, river beds, saline-alkaline lands, eroded hill slopes, mine-spoiled areas and other wastelands. (Muthana and Arora, 1983). Initially, plantations were established mainly for the purpose of conservation, but *P. juliflora* has now become the main source of fuel in rural areas and also to a large extent in urban areas, fulfilling more than 70% of the firewood requirements of rural people in the tropical arid and semi-arid regions of the country. The seed pod is used as a livestock feed, and the wood is also used in charcoal making and occasionally as a timber.

Introduction of other species of *Prosopis* only began after 1960. In 1962, *P. chilensis* and *P. tamarugo* were introduced into Rajasthan and *P. alba* and *P. pubescens* were introduced at CAZRI, Jodhpur in 1966.

These were simple introductory trials to assess the adaptability of these species to the arid environmental conditions of India. A further trial involving six other exotic provenances of *P. juliflora* was established in 1973, and in 1985, under a FAO/IBPGR programme, *P. pallida* was also introduced at CAZRI, Jodhpur. *P. alba*, *P. chilensis*, *P. glandulosa*, *P. hassleri*, *P. nigra*, and *P. pallida* were introduced into the saline-sodic soils of Uttar Pradesh in 1980 and into the saline-sodic soils at Karnal, Haryana in 1985.

In 1991, under a joint Indo-US research project (PL-480), 232 accessions of several *Prosopis* species (*P. alba*, *P. chilensis*, *P. flexuosa*, *P. nigra*, a *Prosopis* hybrid and Peruvian *Prosopis* sp.) were introduced at CAZRI, Jodhpur. Convinced by the excellent performance of these *Prosopis* species in various research stations and their multiple utility, the forest departments of a number of states of India have now started establishing large scale plantation programmes of these species. The following discussion is essentially related to *P. juliflora*, which still dominates the arid zone, although some important aspects of genetic improvement of other introduced *Prosopis* species are included.

Habitat and Growth

The ecological amplitude of *P. juliflora* is very wide, excluding frost-susceptible areas, and it can be grown in any soil type. It can tolerate very severe drought and flourishes even in very low and variable rainfall. It can be grown on the poorest of soils, with low organic matter and nutrient status. In waterlogged and highly saline areas where nothing else can be grown, this species thrives well and produces large quantities of biomass. In the Little Rann and Great Rann of Kutch, Gujarat, it is the predominant species, with plant densities exceeding 2000 individual/ha, depending on the rainfall pattern and available soil moisture. It is an aggressive coloniser in areas where soil moisture availability is adequate. Two forms of *P. juliflora* are observed in its naturalised habitat. The first is the large 'tree' form, with individuals often more than 8 m in height and having a clear single bole up to 3-4 m. Some trees of more than 15 m in height were observed near Jodhpur and in the Allen forest of Kanpur. The second is the small 'bush' form, multi-stemmed at ground level, although occasionally the multi-stemmed nature may be due to coppice regrowth. Though initially bushy by habit, with forking taking place at lower heights, if it is pruned from the base leaving one stem, it can be grown straight. It was also observed that as the tree becomes older, it becomes more thornless, thus it appearing that the presence of thorns in the juvenile stage is a protective mechanism.

Establishing the Plantation

Reproduction and Multiplication

In nature, *P. juliflora* reproduces through self-seeding, with trees producing large quantities of seeds enveloped in indehiscent pods. Pod production varies from tree to tree, from only a few kilograms to 50-70 kg. The pod shape, weight, length and width also vary between species of *Prosopis* and also between provenances of same species (Sharma *et al.*, 1993). Similarly, the seed to pulp ratio also varies considerably. Valdivia (1972) reported that the selection of the most vigorous, healthiest and most productive trees for seed, resulted in highly productive plant types. Carvalho (1963) reported that seeds from well formed, vigorous, thornless and highly productive adult trees should be collected for the improved growth of the progeny. Muthana and Arora (1983) and Mutha *et al.* (1994) also reported that larger and heavier seeds always gave better quality seedlings. However, plantations raised with seed of selected trees exhibited variability over time in growth and subsequent development.

To raise homogeneous plantation stands, multiplication and establishment of cuttings is the only appropriate method, because of the high genetic diversity found in seedlings. Felker and Clark (1981) obtained over 70% rooting with stem cuttings of *P. alba, P. chilensis, P. glandulosa, P. pallida* and *P. velutina*. At CAZRI, Jodhpur, vegetative propagation from stem cuttings of introduced *Prosopis* species gave 40% rooting with *P. alba*, 11% with *P.* sp. 'Peru' and 6% with *P. chilensis* in open nursery conditions. In the controlled conditions of a mist chamber, over 80% rooting was obtained with similar stem cuttings.

Recently, another method of vegetative propagation, that of grafting, has been standardised. Successful wedge grafting was obtained both at the nursery stage with 4 month old seedlings, and with 12 month old field-planted saplings of *P. juliflora*. It was observed that July to early August is the most suitable period for the grafting of 4 month old seedlings, when success rates of greater than 80% were achieved. Wedge grafting in the nursery appears a suitable technology for the production of erect, thornless seedlings for supplying to farmers and other users for direct field planting.

Seed and Nursery Work

The extraction of pure seeds from the pods is very difficult, as each seed is encased in a hard endocarp shell. In nature, the pods are eaten by grazing animals and the seeds pass through the rumen intact, germinating easily on excretion. Natural germination via this process is common, and the rapid and uncontrolled spread of this species can take place. Therefore, for the revegetation of any land with *Prosopis* species in a planned manner, large scale seed extraction is required.

At VRTI (Vivekanand Research and Training Institute, Mandvi, Gujarat), a thresher was developed for seed extraction, where by 1993, 1097 t of *P. juliflora* pods had been collected and processed (Kanzaria and Varshney, this volume).

In the state of Gujarat, more than 10 such threshers are in operation. Some 10-15% of the seeds are damaged during processing, but then no pretreatment is required for germination as the seed coat is scarified by the threshing process. Chemical extraction is also possible, by soaking pod segments in 0.1 % hydrochloric acid solution for 24 h or a 4% solution of sodium hydroxide for at least 30 min (Silva, 1988).

The seeds of *P. juliflora* have a hard seed coat and require pretreatment before sowing. Bohra *et al.* (1994) reported mechanical scarification to be the most effective pretreatment. However, for large quantities of seeds, only sulphuric acid treatment was found to be suitable. Mutha *et al.* (1994) observed that larger and heavier seeds produced higher germination rates and better quality seedlings as compared to smaller seeds.

The seeds, once removed from their endocarps, are used for raising seedlings in the nursery and for direct sowing in the field, however in arid regions, direct sowing of seeds has failed to give any encouraging results, primarily due to low and erratic rainfall. For improved establishment, the raising of seedlings in a nursery is prerequisite. Mutha *et al.* (1995) observed that depth of sowing is very important to obtain high germination. Under nursery conditions, sowing at 1 cm depth gave the maximum germination, while seeds sown at 5 cm depth completely failed to germinate.

In the nursery, seedlings are raised in clear polythene bags or tubes (approximately 11 cm x 4 cm, from 300 gauge plastic). More recently, thermocol pots are also being used for raising seedlings. In Uttar Pradesh and some other states, black polythene bags are being used instead of transparent ones to avoid algal growth, which can occasionally adversely effect seedling development.

Planting

Plant spacing depends upon the objectives of the plantation. If the objective is fuelwood production, then the plant density should be high, with spacings of 2 x 2 m or less. For afforestation programmes for multiple utility, the recommended spacing is 3 x 3 m, especially in the case of saline-sodic soils. For sand dune stabilisation, the spacing used is 5 x 5 m. *P. juliflora* is often used for live fencing and as windbreaks with a recommended spacing of 1 x 1 m. For silvopastoral and other agroforestry systems, the recommended spacing is 10 x 10 m.

Seedlings attain a height of 35-45 cm after four months and are ready for out-planting at selected sites, normally in early July after the first effective rains. At planting, seedlings should be removed from the nursery container, and if the roots are coiled they should be cut or made straight. Pesticides should be applied to the planting hole to prevent termite damage and irrigation should be provided. Weeding should be carried out twice during the monsoon season and once more the following March. Growing saplings should be trained into single-stemmed trees from the very beginning by removing side branches, unless they are grown for the purpose of live fencing.

Production Potential

To determine the optimum age for fuelwood harvesting, a study on yield of *P. juliflora* was carried out at 4 sites; Sardarshahar (268 mm), Jhunjhunu (395 mm), Gadra Road (285 mm), and Bikaner (285 mm). There were wide variations in the fuel yield with respect to age, habitat and the amount of rainfall received. With the increase in rainfall from west to east in western Rajasthan, fuel production showed a corresponding increase. Different workers have reported variable biomass production from different regions. Gurumurti *et al.* (1984) reported 113.25 t/ha when planted at 1.3 x 1.3 m spacing, while Chaturvedi (1985) reported a harvest of only 6.7 t/ha from two year old plantations of *P. juliflora* on saline-sodic (usar) soils. Patel (1986) reported that high density plantations of *P. juliflora* (10000 trees/ha) produced 250 t/ha, 5 years after planting. Rao (1986) recorded a yield of 50 t/ha from Andra Pradesh.

The wood of *P. juliflora* is rated as a high quality timber, although the sapwood is very different from the heartwood. The sapwood is yellow in colour and susceptible to wood boring insects, while the heartwood is a rich dark brown colour and very hard. The grain of the wood is close, straight to slightly interlocking, and medium coarse in texture. The wood is very dimensionally stable and extremely hard, fairly easy to work and finishes smoothly. In Andhra Pradesh alone, 250000 m^3 of *P. juliflora* timber is produced annually (Rao, 1986).

Alternative Uses

Gum production is approximately 40 g/tree/yr. From the Kutch districts of Gujarat, 300 t of gum is being collected annually and is used mainly as an emulsifying agent in industry. Where *P. juliflora* grows abundantly, it is the main source of pollen for honey production, with trees flowering twice a year. Approximately 300 t of honey was collected from *P. juliflora* plantations in the last 5 years in Gujarat, and simultaneously during the process of purification, about 15 t of beeswax was also obtained over the same period. Also, in addition to fuelwood, the GSFDC (Gujarat State Forest Development Corporation) also manufactures high quality charcoal from *P. juliflora* wood in substantial quantities.

P. juliflora also has a large role in soil conservation and improvement. Plantations on salt-affected soils have ameliorating effects, by favourably affecting the physical properties of the soil (Singh *et al.*, 1993). Yadav and Singh (1970) observed that *P. juliflora* plantations on alkaline soils lead to a decrease in the salt content and an increase in organic matter and nitrogen content. Singh *et al.* (1989) reported the effects of *P. juliflora* plantations on improving soil water capacity and moisture relations of alkaline soils, with the depth of infiltration after 18 months after planting being only 2.29 cm in control conditions, and 5.21 cm in the soils where plantations were raised. In Rajasthan, *P. juliflora* and *Acacia tortilis* are the main tree species used for sand dune stabilisation. Plantations of *P. juliflora* are now being established with the use of industrial effluent mixed with water, and also for the revegetation of mine-spoiled areas (Aggarwal *et al.*, 1994; Sharma *et al.*, 1996).

Genetic Improvement

Despite all the advantages of *Prosopis* species, rural people do not favour the planting of *P. juliflora* because of the stiff thorns, and the bushy and invasive nature of the tree. Therefore, efforts are being made to improve the species in order to obtain thornless, straight-boled and high yielding plant types (both for fuelwood and pod production). Sharma *et al.* (1993) selected 44 plus trees of *P. juliflora* from different locations in Rajasthan, Gujarat and Uttar Pradesh, and studied the variability in seed and pod characters, and performance at the nursery and one year old plantation stages, and great variability in the different characters was reported. Kumar *et al.* (1993) also conducted studies aimed at improving the *P. juliflora* growing in saline-sodic soils of Uttar Pradesh. Fifteen provenances were selected from plantations at Fisher forest and Bovine areas of Etawah; Allen Forest, Kanpur; Brindavan block, Mathura; and Kukrail, Lucknow. A very high degree of variability was observed among the different provenances, due to the obligatory out-crossing breeding mechanism of *Prosopis* species. Six provenances of *P. pallida* were introduced in 1985 from Peru and their performance were studied, revealing that accession 1124 had the fastest growth rate, attaining a height of 4.8 m after 6 years (Sharma *et al.*, 1993). Several other studies have been conducted, at CAZRI, Rajasthan (Harsh and Tewari, 1992; 1993), including various species, results from one trial containing 7 species. Results from several other trials conducted at CAZRI, and some others from elsewhere in India are detailed in this volume.

Pod Production

The possibilities for genetic improvement in pod quality and pod production are excellent due to the enormous variation in these factors. For a 5 year old plantation, Felker *et al.* (1984) found that mean pod production per accession ranged from 7.1 kg/tree to 0.0 kg/tree for 25 accessions. Even more striking was that the accession with the greatest mean pod production had a range of 3.2 to 12.2 kg/tree.

Perhaps even more noteworthy than the variation in pod production per tree was the variation in protein and sugar contents. Odoul *et al.* (1986) examined the chemical composition of 30 *Prosopis* accessions grown in 3 environments. He found that the pod sugar composition and pod protein content varied from 11% to 40% and from 9.2% to 18% respectively. Accessions with the high protein and sugar contents were similar at the 3 sites. While heritabilities could not be calculated due to the fact that the male parents were not known, intraclass correlations were 0.40 for pod protein content and 0.30 for pod sugar content over the 3 sites. These values indicated considerable genetic control of these pod characters.

There have been a number of studies that have examined the use of *Prosopis* pods in animal rations (Felker, 1984). *Prosopis* pods were a major food source for native Americans in prehistoric times. This author finds that *Prosopis* flour prepared from the pod mesocarp imparts a pleasant aroma and a much liked flavour to baked pastries, when normal flour is substituted with one third *Prosopis* flour. Saunders *et al.* (1986) described a 1000 kg/hr dry milling facility suitable for processing *Prosopis* pods into a high fibre fraction, a high protein faction, a high sugar fraction and a galactomannam gum fraction. Saunders *et al.* (1986) also described the suitability of the use of *Prosopis* flour into chapatis, crackers, flakes, leavened breads and tortilla chips.

Relative Value of *Prosopis* Wood Products

In designing development projects, it is important to know the relative value of the various uses of *Prosopis* products. The lowest monetary value is that of the direct energy value of *Prosopis* wood. Since there are 17 million BTU per dry ton of *Prosopis* wood, the value of a ton of wood at US$1/million BTU (1 million BTU = 1000 cubic feet of natural gas) is only $17/ton. In contrast, *Prosopis* is much more valuable when converted into dimensional lumber. This is because *Prosopis* wood has exceptional dimensional stability with regard to changing moisture conditions (Tortorelli, 1956; Welden, 1986), and that it has similar hardness and density to oak (*Quercus* spp.) and other exotic tropical hardwoods. When compared with current wholesale prices for cherry (*Prunus avium*), walnut (*Juglans* spp.) and oak lumber at US$1-3 per board foot, the value of *Prosopis* lumber ranges from US$604-1812/dry ton, or US$423-1269/m^3.

This clearly points out the fact that economic development strategies should be geared to production of lumber whenever possible. Since *Prosopis* requires a great deal of pruning to obtain straight poles suitable for lumber, there is a great opportunity for the use of prunings for fuelwood, to be obtained while producing clear boles for high value lumber in long rotations. As pointed out later, we have measured annual growth rates of 1.25-1.50 cm/yr in basal diameter which translates into a 20-30 year rotation for a 40 cm diameter saw log (Cornejo-Oviedo *et al.*, 1991a). We have estimated that these 40 cm basal diameter trees contain about 0.17 m^3 of lumber worth a minimum of US$70/tree (Felker *et al.*, 1988).

While *Prosopis* lumber has exceptional properties it has not been widely used in fine furniture due to the short length and small diameter of the logs. This can be overcome by the fact that there is an increasing tendency to directly produce blanks for furniture parts rather than the standard 8 feet long by 12 inch wide (2.4 x 0.3 m) boards (Araman *et al.*, 1982). A recent survey of hardwood lumber sizes used in the U.S.A. cabinetry industry found that 90% of the actual pieces were less than 60 cm long and less than 15 cm wide (Araman *et al.*, 1982). These authors suggested that standard sized blanks could be produced that are much smaller than normal lumber lengths and widths. These smaller widths and lengths could accommodate smaller sized hardwood logs. This is an approach that would be very feasible with *Prosopis*.

However, the routine unavailability of 2.4 m long *Prosopis* logs precluded its grading and thus marketability through official National Hardwood Lumber Association (NHLA) grading channels. Thus Los Amigos dei Mesquite, the organisation of producers and users of *Prosopis* in the U.S.A., proposed a modified set of grading rules to the NHLA that used shorter and narrower boards, which for the highest grade only needed to be 4 feet (1.2 m) long and 6 inches (0.15 m) wide and have a minimum clear cutting surface free of cracks and defects of 4 inches (0.1 m) by 24 inches (0.6 m). This grading system was presented to the NHLA at their 1993 annual convention in Dallas and was voted on by the membership. Since the NHLA Chief Inspector recommended its approval, it is hoped the general membership will ratify these grading rules. At the NHLA convention, major hardwood buyers from Europe, Japan, Canada, South America and the U.S.A. were present and were all favourably impressed with *Prosopis* lumber. Thus it appears as if the international climate is ripe for an excellent demand to high value *Prosopis* lumber.

Technology in use in north-eastern U.S.A. and Canada for small diameter, short length logs needs to be examined for use in India. This technology is adapted to high volume production from 15-30 cm diameter logs that are less than 1.2 m in length. For example, a 3 man crew has been reported to saw small walnut logs at about 1.5 m^3/hr with a 'bolter saw', and at a value of over US$400/$m^3$ this would be very significant. Small, 2 to 3 person sawmills are commercially available for US$3000-15000, of the bandsaw and circular type. While the circular sawmills take more kerf than the bandsaws, the circular saws are more rugged and can take more abuse with the hard wood of *Prosopis*.

The 2 types of sawmills that seem most adapted to high lumber production from small *Prosopis* logs are the bolter mill and the Scragg mill. The bolter mill typically has one 75 cm diameter blade with a 40 hp motor. The Scragg mill has 2 circular blades, about 75 cm in diameter on the same shaft. For small logs, the spacing between the blades is about 10 to 15 cm. In earlier models, the logs fall into a 'V' through which has the effect of centring them. They are then pushed between the blades by a clip on a chain. This mill cuts a slab off 2 sides of the log at one time to make a 2 sided cant

10-15 cm thick. A Scragg mill at Texas A&M was set to cut 2 logs per minute.

Due to the greater hardness and density of *Prosopis*, cutting tooth angles, horsepower requirements, feed rates and tooth composition must be optimised for *Prosopis*. Mr. S. Lunstrum, sawmill specialist at the Forest Products Laboratory of the U.S. Forest Service in Madison, Wisconsin has a computer program to optimise sawmill systems. Mr. Lunstrum has been most helpful to the Texas A&M University sawmill project. It would certainly be useful if a centre of excellence for *Prosopis* sawmill technology were established in India to support this industry.

Genetic Improvement of *Prosopis*

Due to self-incompatibility in *Prosopis* (Simpson, 1977), the male pollen must come from a different source other than the female parent (mother tree). As a result, there is great genetic variability in *Prosopis*. Hunziker *et al.* (1986) has measured the genetic variability and described numerous natural hybrids in *Prosopis*. Synthetic hybrids have also been created in reciprocal crosses between *P. alba*, *P. velutina*, *P. juliflora* and *P. glandulosa* var. *torreyana*. In addition to obtaining useful hybrids through hybridisation, there is the opportunity to do so by merely choosing outstanding trees and strains from existing genetic material of genetically broad-based field trials. Since *Prosopis* is outcrossed, individual trees do not breed true to type and there is a strong need for asexual propagation.

A review of asexual propagation strategies for *Prosopis* has recently appeared (Felker, 1992). Early work: (1) examined use of a DMSO based rooting formulation (Felker and Clark, 1981); (2) identified optimum temperature, light and photoperiod requirements (Klass *et al.*, 1984); (3) optimal hormone drenches (Klass *et al.*, 1986); (4) fertilisation regimes for stock plants (DeSouza and Felker, 1986) and (5) development of a solar powered mist system for use in remote sites without electrical power (Wojtusik *et al.*, 1994). When optimum environmental conditions of stockplants and cuttings are achieved, almost no hormones are required for rooting. However, when less than optimal environmental conditions are obtained, A DMSO based formulation with 6000 mg/l IBA and 9000 mg/l NAA appears useful in increasing rooting percentages. High biomass producing clones have been obtained from both *P. alba* (Felker *et al.*, 1983a; Felker *et al.*, 1989) and thornless Peruvian *Prosopis* (Wojtusik *et al.*, 1993).

More recently, interspecific graft incompatibility between North American, South American and tropical species has been demonstrated (Wojtusik and Felker, 1993). This will allow genetic improvement of existing weedy, thorny *Prosopis* without having to produce nurseries, transport seedlings and water to the site or transplant and protect the small seedlings. This will also permit utilisation of root stocks specially adapted to edaphic conditions or perhaps to stimulate early flowering or more resistance to cold weather. In the greenhouse, some *Prosopis* have been demonstrated to grow in salinities equivalent to full sea water (Rhodes and Felker, 1987).

Biomass Estimation

Whether one is developing stand management techniques or estimating biomass in plantations, regression equations relating biomass of volume of standing timber to diameter, height or canopy diameter are required. Felker *et al.* (1983b) reported a combined regression based on the harvest of over 1300 trees with ranges in diameter from 0.2 cm to 15.0 cm. Felker *et al.* (1989) reported on regression for larger trees in a 3 year old plantation. El Fadl *et al.* (1989) used a log loader to develop regression equations for 10 entire mature trees. The basal diameter and weights ranged from 20 cm to 55 cm and from 218 kg to 2540 kg. In addition, the small and large diameter and the length of all the individual stem segments were measured. Using the diameter equation developed by Rogers (1984) between log dimensions and lumber recovery, regressions were developed to relate total volume per tree and total lumber per tree to basal diameter.

Management of Existing Stands

While genetic improvement has great potential for *Prosopis*, the immense existing stands of *Prosopis* dictate that better management of the existing stands be conducted. Undeniably dense stands (>10000/ha) of small diameter trees (<3 cm) are undesirable, whereas large trees (>40 cm diameter) on wide spacings (>7 m) are desirable, in providing pods, shade, firewood and enhanced soil fertility. Johnson and Mayeux (1990) found nodules on nearly all *Prosopis* in native field settings in Texas. Generally, the dense stands of small diameter trees are of recent colonisation (within 10 years) and if left to mature would eventually thin out due to self-mortality. The question then becomes, what are the combinations of stem diameter and spacing which are sustainable? Felker *et al.* (1988) measured 27 natural plots ranging in stem density from 18800 stems/ha to 6 stems/ha and observed a negative straight line logarithmic regression equation. This equation predicted that 42 cm diameter trees could be obtained on 9.5 m spacings. Obviously this relation would be sensitive to rainfall and presence of groundwater. The greater the moisture availability, the closer it would be possible to obtain the same large sized trees.

A thinning experiment with dense stands of small trees (Cornejo-Oviedo *et al.*, 1991a) found that stagnated dense stands grew at a rate of only 0.25 cm/yr in basal diameter, whereas stands that were thinned to 10 m spacing and intercropped, grew at a rate of 1.25 cm basal diameter per year. This intercropping also prevented the invasion of dense stands of small trees. Another experiment was conducted with an overstocked stand of large trees (average basal diameter of 16.4 cm) that had a dense shrubby understory, to determine the principle factors determining tree growth (Cornejo-Oviedo *et al.*, 1991b). One treatment consisted of understory removal, one of thinning multi-stemmed trees to single stemmed trees, one of fertilisation with 100 kg P/ha to stimulate nitrogen fixation, and combinations of the treatments above. Despite the fact that the available phosphorus in the soil (Olsens) was only 2 ppm, the phosphorus application had no influence on growth. The only treatments which had a major influence on growth were those that included removal of the dense understory. When these stands were thinned to convert multi-stemmed trees to single stemmed trees, 22 m^3/ha of small diameter firewood was obtained.

The self-thinning line that predicts the spacing required for large trees also suggests that if large trees are eventually obtained, they will provide sufficient intraspecific competition to prevent dense stands of small trees from becoming established. Thus rather than rely on herbicides or mechanical eradication even including bulldozers, it appears as if the most sustainable technique for avoiding dense, weedy stands of small trees is to create large trees on wide spacings that will dominate the site. In this thinning process, it may well be necessary to use selective herbicide treatments to kill individual trees. One of the most effective treatments is a basal application of 175 ml of a trichlopyr plus picloram mixture (Grazon P+D) in 8 l of diesel, that is applied 15 cm above ground level to the stump after cutting.

Thus stand management techniques can be very effectively applied to *Prosopis* to create large, straight, single-boled trees on wide spacings that are useful for lumber, to provide thinnings for fuelwood, to enhance tree growth, and to provide intraspecific competition to prevent the encroachment of dense stands of small trees.

Ecology of Prosopis Juliflora in the Arid Regions of India

Prosopis juliflora was introduced into the Indian sub-continent to Sind province (Pakistan) in 1877 from South America, and later on it was introduced into many dry regions of India (Muthana and Arora, 1983). This species has since become naturalised in the arid and semi-arid parts of the country. With its wide adaptability to arid environments and its drought and disease tolerance, it has virtually exceeded all indigenous species in covering the arid and semi-arid tract. It forms thickets on marginal lands, sandy, rocky and gravely wastelands and alongside roads and railways, and is now one of the major sources of fuelwood to the landless and poor populations. In Tamilnadu, Haryana, Gujarat and Karnataka, farmers raise *P. juliflora* hedges along field boundaries. Its presence all over the area has reduced the threat to important indigenous tree species like *P. cineraria*, *Acacia nilotica* and *A. senegal*, and today it is one of the most important species for desert reclamation (Saxena and Venketeshwarlu, 1991).

Plant Characteristics

P.juliflora (Swartz) D.C. (vernacular names: vilayti, pardesi, babul or mesquite), is a small to moderate sized evergreen tree of 3-4 m height with long drooping branches and spreading crown. This species is largely represented in a shrubby state. There are straight, conical and spinescent stipules of 3-9 mm length. Leaves are bipinnate, leaflets in 13-25 pairs, oblong (3 x 1.7 mm) and dark green. Flowering spikes are axillary, hanging 6-8 cm long, cream to yellow in colour. The pods are highly variable in size and shape, mostly curved or sickle shaped, cream-coloured on ripening and indehiscent. They contain 3-25 ovoid seeds, brown to light chocolate in colour, firmly embedded in the pod.

P. juliflora flowers twice a year, in February-March and August-September, and is a prolific seeder. The pods of autumn flowering mature by May or early June and are dispersed before the onset of the monsoon. The monsoon flowering pods mature from early November to mid December. Summer pods mature over a short period of time, whereas maturity of post-monsoon pods are staggered. In drought years, autumn flowering is very much affected, with trees often failing to flower, but the same plant flowers and fruits subsequently when there is adequate rainfall.

Ecological Attributes

P. juliflora is well adapted to warm and dry tropical climates. It grows well in areas receiving 250-600 mm annual rainfall. It is a fast growing tree and has a deep to very deep, well meshed root system. It is capable of growing on inhospitable habitats such as rocky and saline soils, under adverse climatic conditions. Being drought hardy, disease resistant and having foliage non-palatable to livestock, it does not require any special care for rehabilitating marginal lands or wastelands. Large scale plantation work can be undertaken on such habitats without any fencing. This species is however, quite susceptible to frost, and its distribution is very much restricted in areas where there is a regular occurrence of frost during winter months. such as the interdunal zones of Barmer, Jaisalmer, Bikaner and Churu (all in Rajasthan). In areas which experience frost once every 4 to 8 years, the plant suffers dieback to ground level, and later in February-March the plant sprouts back and assumes its original height within a 9-10 month period.

Major growth takes place after the monsoon, and *P. juliflora* has a great capacity for regeneration and is capable of withstanding drought for long periods of time. In consecutive drought and famine periods, most other plant species fail to withstand the environmental harshness, whereas *P. juliflora* has been recorded to establish and colonise many habitats during such periods due to its aggressiveness and lack of competition.

In low rainfall years or prolonged droughts, growing seedlings survive by folding both the cotyledons and protecting the young leaf. Such hardiness makes it valuable for afforestation and rehabilitation work in arid areas. Studies by Dahl (1982) showed that this species is capable of reducing its transpiration rate drastically and drawing firmly held water from the lower substratum. It is also capable of tolerating water-logging for certain periods.

Distribution

P. juliflora is now found in many states of India; Andhra Pradesh, Karnataka, Maharastra, Orissa, Punjab, Uttar Pradesh, Tamilnadu and West Bengal (Gurumurthi *et al.*, 1984; Patel, 1986), and in almost all districts of Harayana, Gujarat and Rajasthan. In all 11 districts of western Rajasthan, *P. juliflora* has spread widely, with high plant densities encountered in the districts of Pali, Jalore, Jodhpur and Nagaur. There has been a great spread into east and south Rajasthan in the last 25 years. Initially introduced along roads, it has now spread into many habitats, including rocky terrains and saline lands.

In Gujarat, the forest department has undertaken a massive afforestation programme on the coast of Bhavnagar and Kutch districts with great success. High plant densities are encountered in Kutch, where it has fully infested almost all habitats including saline lands. In the Kutch, excessive plant populations of *P.juliflora* have suppressed the growth of native species such as *Salvadora persica*, *Acacia senegal* and *P. cineraria*. Many drainage lines are fully covered by *P. juliflora*, and dense thickets occur in Banni grassland areas which experience high levels of salinity. This extraordinary colonisation of Kutch proves the ability of this species to withstand a high degree of salinity. A span of about 30 years (1962 onward) has converted Kutch into *P. juliflora* woodland.

In districts of Haryana adjacent to Rajasthan, there are also high plant densities of *P. juliflora*, especially on wastelands, roadsides and gravelly areas. All the saline/alkaline wastelands and other marginal lands of Mathura and Agra districts of Uttar Pradesh also have high density thickets. The Delhi region contains a similar situation with widespread stands of *P. juliflora*.

Natural Regeneration

P. juliflora is known to invade vacant sites and establish itself quite successfully. The extensive and quick propagation under natural conditions on various habitats can be attributed to the palatability of its pods to livestock.

Seeds are chemically scarified while passing through the intestines of an animal and are encapsulated in the droppings. Each small faecal pellet may contain 1-3 seeds, and provides the desired moisture and nutrition to the germinating seedling. Even under drought conditions the pellet holds moisture for longer, which enables the seeding to overcome the adverse conditions and establish itself in the first year. These features, and the acclimatisation of saplings, has made it valuable for wasteland development.

Pod Production

There is a high variability in pod size and shape, varying from 8-25 cm in length. Seeds are enclosed in a protective square shaped septum, coated with a sweet, thin, dry pulp, whereas the whole septa body is enclosed in a dry, yellow pulp containing 20-30% sucrose. This makes the pod palatable to livestock, also being low in tannin content (0.74-1.5 %) (Felker, 1982.).

Very little information is available in India about the pod yield per plant or on a per hectare basis. In New Mexico, U.S.A., Garcia (1916) reported 17 kg pods/tree, while Jurriaanse (1973) reported 90-150 kg from 10 year old trees in central Africa. Smith (1953) reported pod yield of 4-20 t/ha from mature *P. juliflora* stands. Studies conducted in Anand gave production at about 19 kg pods/tree/yr (Anonymous, 1981). *P. juliflora* pods contain 13% crude protein, 60% nitrogen free extract, 0.15-0.44% phosphorus and 0.30-0.50% calcium on a dry matter basis (Shukla *et al.*, 1984). Ground pods fed to cattle, did not show any adverse effects, but under uncontrolled feeding the pods gave deleterious effects, resulting in the formation of a compact ball of indigestible pods in the rumen, which caused sickness and even the death of cattle.

Seed Extraction

Each seed is enclosed in a gummy septum which provides resistance to losses from insect attack. A few methods have been developed for the removal of seeds from this septum (Saxena and Khan, 1975). Pods are collected and spread in the sun until they are dry and brittle.

These pods are threshed mechanically into small segments, each generally comprising of one seed encased in its gummy septum. The removal of seeds from the endocarp can then be undertaken by soaking the broken segments in: 0.1 M ammonium chloride solution for 24 h; 95% sulphuric acid for 30 min; 1% solution of sodium hydroxide for 30 min; or in 70-80 °C water for 10-15 min and room temperature water for 24 h. The segments should then be well rinsed with water, and rubbing with a coarse cloth ensures 100% seed removal.

Germination and Establishment

Untreated seed gave only 25% germination, but seed extracted with sodium hydroxide, sulphuric acid or boiled water were also scarified in the process and gave 77-90 % germination. Rapid water uptake of treated seeds was observed. Sulphuric acid treated seeds showed noticeable swelling and began to germinate within 24 h of sowing (Saxena and Khan, 1975).

Most seeds start to germinate in 1-3 days after sowing. The primary root emerges and elongates rapidly for 2-3 days while the cotyledons expand slowly and turn from pale yellow to light green in colour. The cotyledons unfold, and the primary leaf emerges within

3 days of germination. In the first week after germination there is rapid growth in the root rather than the shoot. Seven day old seedlings have a 1:4 shoot:root ratio (Gupta and Balara, 1972). In nature, seedlings have a great capacity to survive despite weed competition.

Once the seedlings grow 5-10 cm above ground level, they have the capacity to survive and endure the hot summer months. The upper portion of the stem develops many dormant buds which can sprout and elongate in a suitable environment. It takes 3 years to initiate flowering and fruiting. *P. juliflora* generally develops a bushy shape with multiple stems in natural stands. If the apical bud is maintained and side branches pruned, then the plant can develop into an erect, well crowned tree.

In young plantations, termite attack in sandy soils may kill seedlings, and can be controlled with a single application of Aldrex (30% Ec @ 0.01%). During drought conditions the desert rat (gerbil) will eat bark for moisture and food, and can kill plants by completely debarking them at ground level, so periodic rat baiting is desirable in the sandy plains.

Biomass Production

Large scale, close-spaced plantations (1.3 x 1.3 m) of *P. juliflora* were planted in Gandhinagar, Gujarat (700 mm m.a.r.), which produced 113 t/ha of biomass dry matter after 4 years (28 t/ha/yr), and the total biomass (root + shoot) produced was 148 t/ha (Gurumurti *et al.* 1984). Fuelwood yield studies on *P.juliflora* plantations conducted by Muthana and Arora (1983) on various habitats and different rainfall zones has shown that 5 year old trees produced 4.3 kg/tree at Bikaner and 1.5 kg/tree at Gadra road (Barmer). At Jhunjhunu, 10 year old trees produced 27.4 kg/tree of fuelwood.

Annual fuelwood production increases with age, however rainfall and habitat factors have a significant impact on production. The ability to coppice well is an important attribute in fuelwood species, to ensure continuous biomass production. Tiwari (1983) reported that *P. juliflora* withstood annual coppicing from the second year of establishment, thus becoming a reliable renewable source of wood. Established plants coppice very well after cutting and assume good growth within 1 year.

Fixation and Inhibition

This genus, like many other leguminous genera, *Acacia*, *Albizia*, *Pongamia* etc., is capable of fixing atmospheric nitrogen.

Jarrell *et al.* (1982) found that the growth of *Prosopis* in Mexico was enhanced by inoculating with effective *Rhizobia*, leading to high productivity. Gupta and Balara (1972) found nodule formation after 6 weeks.

Growth inhibition studies by Lahiri and Gaur (1969) showed that *Prosopis* leaves contain an inhibitor which effects both the process of germination and the growth of seedlings of other species. Large quantities of leaf litter below the tree canopy do not allow other seed to germinate.

Salinity/Alkalinity Tolerance

P. juliflora grows well in areas having high salinity or alkalinity levels, and can tolerate alkalinity as high as pH 9.5 (Singh *et al.*, 1993). It will grow normally in situations where pH is below 9.6 and electrical conductivity (Ec) is below 1.20 ds/m. Similar observations were made in the saline grasslands of Banni (Kutch), where thickets of *P.juliflora* (>2000 trees/ha) have been recorded, on soils with a surface pH of 8.1-8.4 electrical conductivities of between 0.29-0.87 ds/m. It also can continue to grow in a scattered fashion (10-40 trees/ha) with 8.1-9.3 pH and Ec of 4.8-5.91 ds/m. Even in areas of high salinity, it can be planted by amending the soil with farm yard manure and gypsum.

Utilisation

In western Rajasthan, *P. juliflora* is a major source of fodder, fuelwood, charcoal, timber and gums, and has established itself as a species of soil conservation and for aesthetic purposes. It also provides utility to wildlife, where in the arid tracts of Rajasthan and Kutch, ripe fruits of this species have become the main source of nutritive food to wildlife.

Clusters of bushes or thickets provide hiding places for wildlife from hunters and natural enemies. Ripe pods are consumed by livestock in large quantities, and in drought and especially in famine conditions, the pods serve as the main protein source for milk cattle.

Presently 70% or more of the fuelwood demand of people in the arid and semi-arid tract of Rajasthan, Gujarat and Haryana is met by this species. It has a high calorific value and is preferred over indigenous species.

A high quality charcoal is produced, which has virtually replaced *Anogeissus pendula* (Dhokra) as the main charcoal producing species in the arid zone. The timber is hard, heavy,strong, close grained and does not crack. The heartwood is deep brown, while sapwood is thin and yellow, and exists in the ratio of 1:9. Strong and durable timber is obtained from the heartwood, which serves as a valuable material for agricultural implements and household furniture and utensils. The plant produces gum from February to April which can be used for sizing cloth and paper.

Ecology of succession of Prosopis juliflora in the ravines of India

Two *Prosopis* species occur in the ravines of India, the exotic *P. juliflora* and the indigenous *P. cineraria*, both being very important and versatile trees. The ecology of *P. juliflora* has been studied for the past 40 years (Chinnamani, 1992) and is a pioneer species with a very aggressive colonising habit.

The pods and seeds are dispersed by water and wind, but mainly by ruminants, especially goats (Chinnamani and Rathore, 1983). Impenetrable thickets can form in places where there is alkalinity with adequate water. It is constantly being cut as a source of firewood, and it also tolerates severe biotic interference and thrives well even in highly unfavourable situations.

P. juliflora in Ravines of Some Important Indian Rivers

P. juliflora has been introduced for ravine reclamation for the last 40 to 60 years in the shallow, medium and deep ravines of India. These ravines are found mainly on the banks of the rivers Yamuna, Chambal, Beas, Mahi, Banas, Sabarmati, Narmada and Tapi and their tributaries, covering an area of 3.7 million ha. *P. juliflora* has become naturalised in these semi-arid ravinous tracts, regenerating naturally and establishing well at the head, sides and beds of ravines, including precipitous side slopes.

With continuous and rigid protection they tend to allow the natural regeneration of indigenous tree species over time. After 30 to 60 years of rigid protection of stands of *P. juliflora*, the proportion of this species declined from 80-95% to 15-40%, and is gradually replaced by neem (*Azadirachta indica*), sissoo (*Dalbergia sissoo*), babool (*Acacia nilotica*), khair and others, finally becoming one of the associates of the northern tropical thorn forest, the plant composition. In the ultimate dry deciduous forest, *P. juliflora* is completely eliminated in the late successional stages, as can be seen in the ravines of Uttar Pradesh, Rajasthan and Madhya Pradesh.

Ecology and Management of Some Important Species of Prosopis

Prosopis, commonly called mesquite, is distributed in the arid and semi-arid parts of the tropical and sub-tropical regions. Many of the species are now pantropic in their distribution and play an important role in the rehabilitation of wastelands including salt affected lands. A brief account of important species with special reference to their distribution, morphological description, environmental requirements, management and utilisation is presented here.

Distribution

Most of the species are indigenous to North and South America, but also Africa and Asia. Arboreal species such as *P. alba*, *P. chilensis*, *P. hassleri* and *P. nigra* are the largest sub-tropical species found in Argentina, *P. glandulosa* is common from Mexico to Kansas and Louisiana, while *P. pallida* is native to Peru. Many species have been introduced in dry and coastal regions of India, Pakistan, Brazil, Senegal and Sudan, where they have become naturalised and are widely used as a source of fuelwood.

In the Indian sub-continent, *P. juliflora* was introduced into Sind in 1877 and later into many other parts including Hageri, Bellary, Punjab, Agra and Delhi. Now it has spread into a variety of habitats, including saline and alkaline lands, eroded hills and ravines, along river beds, and dry and degraded wastelands, often where rainfall is low and variable. *P. cineraria* is the main native species, found throughout the arid and semi-arid regions, while the shrub *P. farcta*, also native to Asia, has been reported to occur in parts of Punjab and Gujarat. Although less common, *P. glandulosa* is reported to be found introduced in northern India (Maheshwari, 1963; CSIR, 1969). Recently, many other species such as *P. alba*, *P. articulata*, *P. glandulosa*, *P. laevigata*, *P. tamarugo* and *P. velutina* from many provenances have been introduced into India particularly in dry regions and wastelands.

Morphology and Taxonomy

Prosopis Linn. (family Leguminosae, subfamily Mimosoideae) is a genus of generally fast growing, mostly thorny shrubs and trees, having very wide morphological variations and an ability to withstand wide climatic and edaphic diversity. These hardy and fast growing trees also have a remarkable coppicing ability. They have been seen as excellent plants for afforestation programmes, especially in India, because of their ability to grow in dry and degraded lands, having extensive lateral root systems and long taproots (penetrating up to 24 m), resistance to grazing, low cost of production and good quality firewood. The following is a brief description of the species which are grown in India, introduced, naturalised or indigenous.

P. alba Grisebach is a round crowned tree, (introduced into India from the U.S.A, but of South American origin), 5-15 m tall, sometimes with a trunk as large as 1 m in diameter and a short bole with many branches. The sapwood is yellowish and the heartwood dark brown. It is a thorny tree, but thornless varieties are also found. The leaves are bipinnate, with rachis up to 5 cm long, bearing long leaflets, 1-2 mm wide. The foliage is pinkish on young branches, developing a bluish tinge when mature. Flowers are abundant, greenish-white, on spike-like racemes. The pods are beige to off-white (from where the species name originates), typically 18-22 cm long, 4-5 mm thick and sickle-shaped. In contrast, other Argentinean species have red-tinged to dark purple pods (e.g. *P. flexuosa* and *P. nigra*).

P. articulata S. Watson has rachis up to 3 cm long, bearing two pairs of pinnae (7-9 mm long) with 30-50 pairs of leaflets per pinnae (6-8 mm long and 1 mm wide). Foliage and branches have a bluish tinge when young.

P. chilensis (Molina) Stuntz emend. Burkart is a small to medium sized tree up to 20 m in height and a bole up to 1 m in diameter. Trees are fast growing, spiny or sometimes unarmed, semi-evergreen to evergreen, large-crowned with drooping branches. The bark is greyish or dark brown, fissured, exuding gum. Leaves are dark green, bipinnate with 2-4 pairs of pinnae, with 10-48 pairs of leaflets, each leaflet 5-20 mm long. Flowers are small, greenish-yellow, numerous, in dense spikes. Pods are straight, falcate, flat or cylindrical (10-25 cm x 8-15 mm), containing 10-30 seeds being ovoid, flattened, hard, yellowish-brown, shiny and 7 x 3 mm in size (there being 9800-36000 seeds/kg). Germination is 50-100% following scarification in sulphuric acid or hot water. This species has been previously wrongly identified as *P. juliflora* or one of its varieties, but is easily distinguished by its longer and wider spaced leaflets and red-tinged seed pods.

P. farcta (Solander ex Russell) MacBride (synonym *P. stephaniana* Kunth ex Spreng.) is native, being a spiny bushy shrub or small tree, with downy, slender whitish branches, and armed with copious slender stramineous prickles. Leaflets similar to *P. cineraria* but smaller and downier, and with larger flowers. Pods are oblique and oblong, 1.5-3.0 cm long, 8-10 mm thick, black, rugose, obtuse, and contracted abruptly into the stalk, occasionally with galls.

P. glandulosa Torrey (synonym *P. juliflora* D.C. var. *glandulosa* Cock) is a shrub or small tree, 3-7 m tall with straggling or pendulous branches, stout spines. The tree has 1-2 pairs of pinnae each with 10-18 pairs of leaflets, rather distant along the rachis, linear and oblong in shape, 6-8 mm long and 1 mm broad. Flowers are yellow, in axillary spikes, with pods 13-20 cm long and 5-10 mm in diameter, linear, straight or falcate. It has also been wrongly identified as *P. juliflora*, but is distinguished from it by being deciduous, having falcate pods and larger leaflets without a mid rib. Two varieties are recognised, var. *torreyana* growing primarily in the deserts of the south western U.S.A. and northern Mexico (Hilu *et al.*, 1982), and var. *glandulosa* found from Mexico north to Kansas and east to Louisiana (Burkart, 1976).

P. laevigata (Humboldt & Bonpland ex Willdenow) M.C.Johnston has been introduced in Karnal, Haryana. Native to Mexico, it has 1-2 pairs of pinnae, each 2-3 cm long and having 12-18 pairs of leaflets 3-5 mm long.

P. juliflora Swartz D.C. is the most common species in India. It is a fast growing, thorny, drought resistant, evergreen, large crowned, deep rooted bush or tree up to 10 m or more in height, with drooping branches. There are usually 4 pinnae with 17-20 pairs of dark green leaflets, obliquely oblong in shape. Flowers yellowy green, while pods are straw coloured when ripe, 12-25 cm long, curved, with brown seeds.

P. pallida (Humboldt & Bonpland ex Willdenow) H.B.K. (synonym *P. limensis* Bentham), native to Peru, Colombia and Ecuador, is a scraggly tree, thriving well in hot and saline regions. It grows to 8-20 m in height with a trunk of 60-80 cm in diameter, and branches reminiscent of twisted rope. On poor sites it is shrubby, resembling *P. juliflora.* It has sharp spines and pale-yellow flowers, with greyish green leaves and abundant pods 12-24 cm long, but thornless trees are also found.

P. tamarugo F. Philippi, native to the Atacama desert in Chile, is a long lived tree up to 18 m tall, extremely drought resistant and salt tolerant, with a deciduous, open crown. It has golden yellow flowers on long cylindrical spikes, with curved pods containing about 7 seeds embedded in a brownish pulp. It produces a dense mat of lateral roots and a quick growing tap root.

P. velutina Wooton (synonyms *P. chilensis* var. *velutina* Standley, *P. juliflora* var. *velutina* Sarg.) grows up to 16 m in height and reaches a bole diameter of 60 cm, with usually 2 pairs of pinnae, 4-5 cm long, having 18-24 pairs of leaflets. Leaves and inflorescence are densely tomentose.

Environmental Requirements

Most of the species are drought tolerant and suitable for planting in arid and semi-arid regions in a variety of soil conditions. *P. juliflora* can be found growing in many environments such as mine spoils, along roadsides, railway tracks, alkali and saline tracts and on rocky terrain. It is predominant in the extremely saline soils in the Rann of Kutch, Gujarat, where most other species, except some true halophytes, fail to survive. It has been found growing in very warm, dry climates and tolerates high sodicity (pH up to 10 or more) and salinity (electrical conductivity 44 dS/m or more). *P. chilensis* is equally salt tolerant, even tolerating sea-water salinities. It is native to regions experiencing lower winter temperatures and low rainfall with ground water between 3 and 10 m deep. *P. alba* is native to high plains up to 1500 m elevation and tolerates high sodicity (pH 10).

P. pallida is a frost sensitive species, native to the incredibly dry, saline coastlines of Peru and Ecuador, remaining shrubby on poor sites. *P. glandulosa* grows in a wide range of soils and is moderately salt and frost tolerant. It thrives in very high temperatures (>38°C) and in areas with low rainfall, but usually in areas with accessible ground water reserves. *P. tamarugo* tolerates a wide temperature range (12-36°C) and is extremely drought resistant and salt-tolerant. It is found in areas with thick, surface salt-encrustations.

Many species can grow in soil salinities equivalent to sea water, and Felker *et al.*, 1981) found that all species tested could tolerate salt concentrations of 6,000 mg/l with no reduction in growth. *P. velutina, P. pallida* and *P. tamarugo* could grow in salinity levels of 18,000-36,000 mg/l. These species are reported to be capable of fixing nitrogen, thus having potential as nitrogen fixing halophytes. Characteristics of 5 important species.

Propagation and Silviculture

Prosopis seeds are not easily removed from the gummy pulp. The pods can be processed in a grinder after drying at about 50°C overnight, which will also scarify the seeds. Germination pretreatments include mechanical scarification, soaking in concentrated sulphuric acid for

20 min or a 20% solution for 60 min, or immersed in boiling water and allowed to cool for 24 h. There are 25,000-36,000 seeds/kg. Outstanding plus trees of *P. alba* have been cloned by vegetatively propagating shoot cuttings requiring the control of light intensity and air temperature (Felker, 1991).

Plant spacing varies, for energy plantations being 1 x 1 m or 2 x 2 m, but for a silvopastoral system 2 x 4 m is recommended, which may increase to 4 x 4 m for arable crops 3 years after planting. Soil with compacted hard pan layers, as in many alkali soils of the Gangetic plains, should have a hole dug with a tractor-mounted auger, 20-25 cm diameter and 1-1.5 m deep and the holed refilled with the original soil with 8 kg manure, 3 kg gypsum 10 g zinc sulphate and some insecticide (Dagar and Singh, 1994). Trees respond well to irrigation, with excellent coppicing ability, and maximum fodder yields are obtained when the trees are pollarded on a three year rotation.

The best growth is achieved in areas when the root system can reach groundwater, but in low rainfall areas, especially in fast draining soils, irrigation may be required during establishment. Early pruning to encourage an erect form is recommended.

Biomass Production

In India, *P. juliflora* grows rapidly on all kinds of soil, with mean annual increments of 66 cm in height, 2.51 cm in collar diameter and 1.73 cm in breast height diameter recorded, and a total yield of 50-60 t/ha after 10 years and 75-100 t/ha after 15 years (Vimal and Tyagi, 1986). In semi-arid Gandhinagar in Gujarat, a 4 year old plantation (1.3 x 1.2 m spacing) produced 148.63 t/ha dry biomass (Gurumurti *et al.*, 1984). Studying the response of 3 planting techniques, trench (30 x 30 cm deep), pit (30 x 30 x 30 cm), and auger hole (15 x 90 cm deep), the lopped harvested biomass was 34.3, 36.0 and 42.6 t/ha respectively. In response to the soil amendments of gypsum, gypsum + rice husk, and gypsum + farm yard manure, a total biomass of 34.3, 49.5 and 52.3, t/ha, respectively was obtained, compared with only 3.1 t/ha without any amendments, in highly alkali soils (pH 10.4).

Bibliography

Anderson, E. F.: *The Cactus Family*, Timber Press, Portland, Oregon, 2001.

Andrews L.: *Citrus Production - Orange*, St. Augustine, Trinidad and Tobago, 1990

Arora, Chandralekha : *Advances in Horticulture Science Research*, Oxford Book Company, Delhi, 2009.

Benson, L. H.: *The Cacti of the United States and Canada*. Stanford University Press, Stanford, California, 2001.

Benu Singh: *A Modern Book on Forestry and Horticulture*, Vista International Pub, Delhi, 2010.

Bhandari, M.M.: *The Flora of the Indian Desert*. Scientific Publisher, Jodhpur, 1990.

Bhattacharjee, Supriya Kumar: *Advances in Ornamental Horticulture*, Pointer, Delhi, 2006.

Bhutani, R.C.: *Fruit and Vegetable Preservation*, Biotech Books, Delhi, 2003.

Chadha, K.L. : *Advances in Horticulture*, Malhotra Publishing House, Delhi, 2006.

Collymore L.: *Fruit Production in Barbados*, Port of Spain, Trinidad and Tobago, 1996.

Dabholkar, A.R. : *General Plant Breeding*, Concept, Delhi, 2006.

Doijode S. D.: *Seed Germination in Fruits*, New Delhi, Malhotra Publishers, 1993.

Goswami, C L : *Plant Physiology : Research Methods*, Scientific, Delhi, 2007.

Julia F.: *Fruits of Warm Climates*, Miami, Julia F. Morton Publisher, 1987.

Kumar Supriya : *Amenity Horticulture, Biotechnology and Post-Harvest Technology*, Pointer, Delhi, 2006.

Little, E.L.: *Common Fuelwood Crops*, Communi-Tech Assoc., Morgantown, West Virginia, 1983.

Malins A.: *Postharvest Handling of Pineapple and Mango*, Port of Spain, Trinidad and Tobago, 1992.

Mitra S.: *Postharvest Physiology and Storage of Tropical and Subtropical Fruits*, Oxon, CABI, 1997.

Morton, Julia F.: *Fruits of Warm Climates*. Wipf and Stock Publishers, 1987.

Pareek O. P.: *Advances in Horticulture: Fruit Crops,* New Delhi, Malhotra Publishing House, 1993.

Paul B. : *Food Biotechnology in Ethical Perspective*, Aspen, CO: Aspen Publishers, 1997.

Peter, K.V. : *Basics of Horticulture*, New India, Delhi, 2009.

Popenoe, W.: *Manual of Tropical and Subtropical Fruits*. Hafner Press, 1974.

Proctor F. J.: *Onions in Tropical Regions*, Kent, Natural Resources Institute, 1990.

Purohit, S.S. : *Organic Farming : Biocontrol and Biopesticide Technology*, Agrobios, Delhi, 2008.

Saini, M.L. : *Plant Breeding and Crop Improvement*, CBS, Delhi, 1997.

Saroj, P.L. : *Advances in Arid Horticulture*, International Book Dist, Delhi, 2004.

Sharma, V.K. : *Advances in Horticulture : Strategies, Production, Plant Protection and Value Addition*, Deep and Deep, Delhi, 2004.

Singh Z.: *Propagation and Rootstocks of Citrus*, New Delhi, Malhotra Publishers, 1993.

Thomas E.: *Fruit Production in St. Kitts and Nevis*, Port of Spain, IICA, 1996.

Index

F

G

H

I

K

M

N

T

V

W

❑❑❑